Remote Sensing Image Fusion

Signal and Image Processing of Earth Observations Series

Series Editor

C.H. Chen

Published Titles

Remote Sensing Image Fusion

Luciano Alparone • Bruno Aiazzi
Stefano Baronti • Andrea Garzelli

CRC Press
Taylor & Francis Group
Boca Raton London New York

CRC Press is an imprint of the
Taylor & Francis Group, an **informa** business

CRC Press
Taylor & Francis Group
6000 Broken Sound Parkway NW, Suite 300
Boca Raton, FL 33487-2742

First issued in paperback 2019

© 2015 by Taylor & Francis Group, LLC
CRC Press is an imprint of Taylor & Francis Group, an Informa business

No claim to original U.S. Government works

ISBN-13: 978-1-4665-8749-6 (hbk)
ISBN-13: 978-0-367-86818-5 (pbk)

Visit the Taylor & Francis Web site at
http://www.taylorandfrancis.com

and the CRC Press Web site at
http://www.crcpress.com

Contents

List of Figures

List of Tables

Foreword

We are delighted to introduce the first book on Remote Sensing Image Fusion. In most emerging research fields, a book can play a significant role in bringing some maturity to the field. Research fields advance through research papers. In research papers, however, only a limited perspective can be provided about the field, its application potential, and the techniques required and already developed in the field. A book gives such a chance. We liked the idea that there will be a book that will try to unify the field by bringing in disparate topics already available in several papers that are not easy to find and understand. The book is a joint effort of four researchers who have been active in remote sensing image processing for almost two decades.

Rather than a historical review of the main achievements in Remote Sensing Image Fusion, this book brings a new perspective on the topic. The development of newer and newer processing methodologies have been substantially motivated by progress in sensor technologies of imaging systems and their widespread coverage of wavelengths, from the visible spectrum to microwaves, as well as by submetric ground resolutions, so far inconceivable for civilian applications.

The common guideline of the book is that relatively simple processing methodologies that are tailored to the specific features of the images may be far winning in terms of reliable performances over more complex algorithms that are based on mathematical theories and models unconstrained from the physical behaviors of the instruments under concern. Under this perspective, quality assessment plays a key role in determining more and more performing fusion methods. In a first generation of image fusion techniques, quality issues were at an early stage and substantially disregarded the requirement of high-quality synthesis of spectral information, which is the guideline of second-generation techniques pursued nowadays by a majority of researchers.

In parallel, after the second generation has reached its maturity, a third generation has started flourishing. Such techniques are more or less exploiting new concepts and paradigms of information science and of signal processing and are characterized, at least thus far, by massive computation requirements, presently not affordable for real size images, and performances comparable to the best ones achieved by the state of the art of the second generation. Though third-generation fusion techniques have polarized the interest of researchers over the last five years, the full maturity is still relatively far to come. Only the lessons learned with second-generation methods will be capable of fostering the excellence among the myriad of methods that are proposed almost every day by the scientific literature.

This book is expected to bring a new perspective to a multidisciplinary research field that is becoming increasingly articulate and comprehensive. The leading effort is fostering signal/image processing methodologies toward the goal of information extraction, either by humans or by machines, from remotely sensed images. However, the book may also be useful to newcomers, either students aiming to improve their specialization in environmental sciences or users, who wish to improve their professional skills beyond the routine application of an established practice.

Preface

Nowadays, approximately four TB of image data are collected daily by instruments mounted on satellite platforms, not to mention the data produced by a myriad of specific campaigns carried out through airborne instruments. Very high-resolution (VHR) multispectral scanners, IKONOS, QuickBird, GeoEye, WorldView, Pléiades, just to mention the most popular, and especially the related monospectral panchromatic instruments are responsible for a large part of the amount. Imaging spectrometers with tens to hundreds of bands will significantly contribute after the launch of the upcoming PRISMA and EnMap missions. In parallel, synthetic aperture radar (SAR) satellite constellation systems, TerraSAR-X/Tandem-X, COSMO-SkyMed, RadarSat-2, and the upcoming RadarSat-3 and Sentinel-2 are taking high-resolution microwave images of the Earth with ever improved temporal repetition capabilities.

The availability of image data with spectral diversity (visible, near infrared, short wave infrared, thermal infrared, X- and C-band microwaves with related polarizations) and complementary spectral-spatial resolution, together with the peculiar characteristics of each image set, have fostered the development of fusion techniques specifically tailored to remotely sensed images of the Earth. Fusion aims at producing an extra value with respect to those separately available from the individual datasets. Though the results of fusion are more often analyzed by human experts to solve specific tasks (detection of landslides, flooded and burned areas, just to mention a few examples), partially supervised and also fully automated systems, most notably thematic classifiers, have started benefiting from fused images instead of separate datasets.

The most prominent fusion methodology specifically designed for remote sensing images is the so-called panchromatic sharpening or *pansharpening*, which in general requires the presence of a broadband in the visible or visible near-infrared (V-NIR) wavelengths, with ground resolution that is two to six times greater than that of the narrow spectral bands. Multispectral pansharpening can be brought back to the launch of the first SPOT instrument, which was first equipped with a panchromatic scanner together with the multispectral one. Recently, hyperspectral pansharpening has also started being investigated by an increasing number of scientists, with the goal, for example, of coupling spatial and spectral detection capabilities in a unique image.

The fusion of images from heterogeneous datasets, that is, of images produced by independent modalities that do not share either wavelengths or imaging mechanisms is a further task, which is pursued in remote sensing applications. A notable example is the sharpening of thermal bands through simultaneous visible near-infrared observations. This issue has also been largely addressed outside the scope of remote sensing, for detection applications, both military and civilian.

A more unconventional and specific topic is the fusion of optical and SAR image data. Several fusion methods, in which an optical image is enhanced by a SAR image, have been developed for specific applications. The all-weather acquisition capability of SAR systems, however, makes a product with opposite features to appear even more attractive. The availability of a SAR image spectrally and/or spatially enhanced by an optical observation with possible spectral diversity is invaluable, especially because the optical data may not always be available, particularly in the presence of cloud covers, while SAR platforms capture images regardless of sunlight and meteorological conditions.

Symbol Description

$\boldsymbol{\alpha}$	Sparse vector in compressed sensing pansharpening.
$\tilde{\mathbf{A}}_k$	Interpolated aliasing pattern of the kth MS band.
$\tilde{\mathbf{A}}_P$	Interpolated aliasing pattern of Pan.
β	Multiplicative factor for the CSA method.
$\mathbf{B}_k$	Original kth image band.
$\tilde{\mathbf{B}}_k$	kth image band interpolated to Pan scale.
$\hat{B}_k$	Pansharpened kth image band.
δ	Spatial detail image.
$\delta(\Delta x, \Delta y)$	Spatial detail image in the presence of Pan-MS misalignment.
D	Dictionary matrix in compressed sensing pansharpening.
D_λ	Spectral distortion.
D_s	Spatial distortion.
Δx	Horizontal spatial misalignment between Pan and interpolated MS.
Δy	Vertical spatial misalignment between Pan and interpolated MS.
$\mathbf{D}_x^{(I)}$	Derivative of $\mathbf{I}$ in the x direction.
$\mathbf{D}_y^{(I)}$	Derivative of $\mathbf{I}$ in the y direction.
$\mathbf{D}_x^{(k)}$	Derivative of $\tilde{\mathbf{M}}_k$ in the x direction.
$\mathbf{D}_y^{(k)}$	Derivative of $\tilde{\mathbf{M}}_k$ in the y direction.
$\mathbf{F}$	Sparse matrix in restoration-based pansharpening.
$\mathbf{g}^*$	Highpass filter impulse response for GLP schemes.
g_k	Global injection gain of the kth band.
$\mathbf{G}_k$	Space-varying local injection gain of the kth band.
h_j^*	Equivalent lowpass impulse response of the jth level of ATW.
H_j^*	Equivalent lowpass frequency response of the jth level of ATW.
H_k	PSF for the kth band in restoration-based pansharpening.
$\mathbf{I}$	Intensity image.
$\mathbf{I}(\Delta x, \Delta y)$	Intensity image in the presence of Pan-MS misalignment.
$\mathbf{I}_k$	Intensity image for the kth band in the BDSD method.
K	Number of bands.
$\tilde{\mathbf{M}}_k$	Original kth MS band expanded to the Pan scale.
$\hat{\mathbf{M}}_k$	Pansharpened kth MS band.
$\hat{\mathbf{M}}_k(\Delta x, \Delta y)$	Pansharpened kth MS band in the presence of Pan-MS misalignment.
$\tilde{\mathbf{M}}_k^*$	Interpolated aliasing free kth MS band.
$\hat{\mathbf{M}}_k^*$	Pansharpened kth band obtained from aliasing free Pan and MS.
p	Integer reduction factor for GLP.
$\mathbf{P}$	Panchromatic image.

q	Integer expansion factor for GLP.
Q	Universal Image Quality Index.
$Q2^n$	Generalization of $Q4$ to 2^n-band images.
$Q4$	Extension of the Universal Image Quality Index to 4-band images.
$\mathbf{P}_L$	Lowpass-filtered panchromatic image.
$\mathbf{P}_L^*$	Lowpass-filtered Pan downsampled by r and then interpolated by r.
r	Spatial sampling ratio between original MS and Pan.
θ	Soft threshold of modulating SAR texture.
$\mathbf{t}$	Texture image extracted from SAR for optical-SAR fusion.
$\mathbf{t}_\theta$	Texture image $\mathbf{t}$ thresholded by θ.
$\mathbf{T}_k$	kth TIR band.
$\widetilde{\mathbf{T}}_k$	kth TIR band interpolated to the scale of the V-NIR band.
$\widehat{\mathbf{T}}_k$	Spatially enhanced kth TIR band.
$\mathbf{V}$	Enhancing V-NIR band.
$\mathbf{V}_L$	Lowpass-filtered version of $\mathbf{V}$.
$\mathbf{V}_L^*$	Lowpass-filtered $\mathbf{V}$ downsampled by r and then interpolated by r.
w_k	Spectral weight of the kth band for intensity computation.
$w_{k,l}$	Spectral weight of the lth band for $\mathbf{I}_k$ computation.

List of Acronyms

ADC	Analog-to-Digital Converter
ALI	Advanced Land Imager
ASI	Agenzia Spaziale Italiana
ASC-CSA	Agence Spatiale Canadienne-Canadian Space Agency
ATW	À Trous Wavelet
AVHRR	Advanced Very High Resolution Radiometer
BDSD	Band Dependent Spatial Detail
CC	Correlation Coefficient
CCD	Charge-Coupled Device
CNES	Centre Nationale d'Etudes Spatiales
CST	Compressed Sensing Theory
COSMO	COnstellation of small Satellites for Mediterranean basin Observation
CS	Component Substitution
CSA	Constrained Spectral Angle
DEM	Digital Elevation Model
DFB	Directional Filter Bank
DLR	Deutsches zentrum für Luft- und Raumfahrt
DTC	Dual-Tree Complex
DWT	Discrete Wavelet Transform
EHR	Extremely High Resolution
ELP	Enhanced Laplacian Pyramid
ELR	Extremely Low Resolution
EMR	ElectroMagnetic Radiation
EnMAP	Environmental Monitoring and Analysis Program
ENVI	ENvironment for Visualizing Images
EnviSat	Environmental Satellite
ERGAS	Erreur Relative Globale Adimensionnelle de Synthèse
ERS	European Remote-Sensing satellite
ESA	European Space Agency
FOV	Field Of View
GGP	Generalized Gaussian Pyramid
GIHS	Generalized Intensity-Hue-Saturation
GLP	Generalized Laplacian Pyramid
GP	Gaussian Pyramid
GPS	Global Positioning System
GS	Gram-Schmidt

HPF	High Pass Filtering
HPM	High Pass Modulation
HS	HyperSpectral
IFOV	Instantaneous Field Of View
IHS	Intensity-Hue-Saturation
IMU	Inertial Measurement Unit
JERS	Japanese Earth Resources Satellite
JPL	Jet Propulsion Laboratory
KoMPSat	Korean Multi Purpose Satellite
LASER	Light Amplification by Stimulated Emission of Radiation
LiDAR	Light Detection And Ranging
LP	Laplacian Pyramid
LWIR	Long Wave InfraRed
MeR	Medium Resolution
MMSE	Minimum Mean Square Error
MoR	Moderate Resolution
MRA	MultiResolution Analysis
MS	MultiSpectral
MSE	Mean Square Error
MTF	Modulation Transfer Function
MWIR	Medium Wave InfraRed
NASA	National Aeronautics and Space Administration
NDVI	Normalized Differential Vegetation Index
NIR	Near InfraRed
NN	Nearest Neighbor
NSCT	NonSubsampled Contourlet Transform
NSDFB	NonSubsampled Directional Filter Bank
NSP	NonSubsampled Pyramid
OLI	Operational Land Imager
OTF	Optical Transfer Function
PCA	Principal Component Analysis
PR	Perfect Reconstruction
PRF	Pulse Repetition Frequency
PRISMA	PRecursore IperSpettrale della Missione Applicativa
PSF	Point Spread Function
QFB	Quincunx Filter Bank
QMF	Quadrature Mirror Filter
QNR	Quality with No Reference
RCM	RadarSat Constellation Mission
RMSE	Root Mean Square Error
RS	Remote Sensing
SAM	Spectral Angle Mapper
SAR	Synthetic Aperture Radar
SHALOM	Spaceborne Hyperspectral Applicative Land and Ocean Mission
SIR	Shuttle Imaging Radar

SLAR	Side-Looking Airborne Radar
SNR	Signal-to-Noise Ratio
SPOT	Satellite Pour l'Observation de la Terre
SRTM	Shuttle Radar Topography Mission
SS	SuperSpectral
SSI	Spatial Sampling Interval
SWIR	Short Wave InfraRed
SWT	Stationary Wavelet Transform
TDI	Time Delay Integration
TIR	Thermal InfraRed
UDWT	Undecimated Discrete Wavelet Transform
UIQI	Universal Image Quality Index
UNBPS	University of New Brunswick PanSharp
US	UltraSpectral
VIR	Visible InfraRed
VIRS	Visible and InfraRed Scanner
VHR	Very High Resolution
V-NIR	Visible Near-InfraRed
WPD	Wavelet Packet Decomposition

Chapter 1

Instructions for Use

1.1 Introduction

Remote sensing (RS) image fusion is a relatively new discipline lying within the generic framework of data fusion. The common perspective is remote sensing, regarded as the ensemble of tools and techniques aimed at extracting information on the Earth's surface and the overlying . Thus, RS image fusion is the synergic combination of two or more image datasets, aimed at producing a knowledge of the phenomena under investigation better than the knowledge achievable from individual datasets.

The mingling of expertises from disparate fields, like optoelectronic, microwave technology, signal processing, computer science, statistical estimation theory, on one side, and environmental monitoring with all related disciplines, on the other side, makes RS image fusion a unique opportunity of fostering existing methodologies and technologies toward a common objective: the safeguard of our planet from a number of risks and hazards that threaten humans' existence, thereby affecting the quality of their lives and their expected lifetime.

RS image fusion has been traditionally pursued by specialists in remote sensing, mainly with the specific goal of expediting visual analysis carried out by human experts. Only recently the possibilities of performing automated analyses on fusion product datasets has started being investigated. Thanks to cross-fertilizations from the signal processing and image analysis fields, RS image fusion has now reached the status of a standalone discipline, rather than of an ensemble of more or less empirical techniques.

The demand of RS image fusion products is continuously growing, as it is witnessed by the increasing diffusion of commercial products using very high-resolution (VHR) multispectral images. As examples, Google Earth and Microsoft Virtual Earth exploit *pansharpening* fusion to yield VHR observations of the Earth, wherever possible.

The different coverage capabilities and imaging mechanisms of optoelectronic (multispectral scanners) and microwave (synthetic aperture radar) instruments makes fusion of such heterogeneous datasets extremely attractive to solve specific monitoring problems in the most disparate environmental and atmospheric conditions.

1.2   Aim and Scope of the Book

This volume features the unique asset of being unconstrained from earlier textbooks on the same topic. In earlier RS books, at most one chapter was devoted to fusion, which was perceived as a new discipline, transversal between applications and methodological sciences:

(a) geology, hydrology, agronomy, physics;

(b) signal and image processing, computer science, mathematics.

On the other side, excellent textbooks on data fusion were more oriented toward the *ontology* of fusion, pertaining the most disparate applications, many of which being of military interest, and datasets (images, signals, point measurements), rather than providing a review of basic and advanced methodologies aimed at the synergy use of RS imagery of the Earth to solve specific problems in civilian contexts.

This book deals with image fusion methods specifically designed for RS imagery, through a comprehensive classification and a rigorous mathematical description of advanced and state-of-the-art methods for pansharpening of multispectral images, fusion of hyperspectral and panchromatic images, fusion of data from heterogeneous sensors, like optical and synthetic aperture radar (SAR) images and integration of thermal and visible/near-infrared images. Fusion methods exploiting new trends of signal/image processing, like compressed, a.k.a. compressive, sensing and sparse signal representations, will be addressed in the near-to-final chapter.

This volume represents a synthesis of an over 10-year activity of the authors in the specific field. Its key assets are:

1. a synoptic and updated overview of the major imaging modalities and of the related instruments and data products;

2. a critical review of the literature over the last two decades;

3. a parallel perspective between progresses in instrument technologies and in data fusion methods;

4. a unifying framework in which the majority of existing fusion methods can be accommodated and new methods can be designed and optimized;

5. specific attention toward solutions implemented in commercial products and software packages over the last years;

6. a projection of new paradigms of signal / image processing, like compressed sensing and sparse representation theory, toward RS image fusion.

For RS image analysis and classification, an excellent and successful textbook, which has reached the fifth edition, is [221]. To the best of the authors' knowledge, no competitive books focus on RS image fusion, some books lacking a deep insight into mathematical issues, others addressing a wide range of applications far beyond remote sensing.

This book is intended for an international audience. The problem of merging image data with different spectral and spatial resolutions or acquired by heterogeneous sensors is of primary interest in all continents: existing space missions and future RS systems are planned in Europe, America, and Asia by national and transnational space agencies and private companies.

Though intended as a textbook for MSc and postgraduate courses, the book may be useful to researchers and scientists in RS and related disciplines. The authors highlight that strictly non-multimedia applications and related datasets are pursued intentionally. This issue is strengthened by the totally different definitions of the term *quality* from the multimedia world.

1.3 Organization of Contents

This volume is organized into 12 chapters covering the majority of topics related to RS image fusion for Earth observation.

1. **Instructions for Use.**
   The concept of fusion for RS image data is introduced. Aim and scope of the book is explained. Organization of contents is outlined and the main content of each chapter is briefly reviewed.

2. **Sensors and Image Data Products.**
   The fundamental physical laws underlying remote sensing are briefly reviewed, together with a synoptic definition of frequency bands and wavelength intervals of the various systems. A non-exhaustive review of the most relevant spaceborne imaging instruments (optical, thermal, microwave), and related data products over the last decades is performed. Increasing trends in spatial, spectral, radiometric and temporal resolution, whenever applicable, are shown to motivate the birth and development of fusion methods and their ongoing progresses.

3. **Quality Assessment of Fusion.**
A critical survey of earlier definitions of quality of fusion products is performed. After that, a categorization of more recent and developed definitions based on the *consistency* of the fusion product to its components, according to the properties that are inherited, is attempted. Presentation of the main protocols of quality evaluation and related drawbacks constitutes the core of the chapter, together with the review of selected similarity / dissimilarity indices, for both monoband and multiband images, to be used for verifying consistency according to the various protocols. Ultimately, hints on the possible extension of quality assessment of multispectral pansharpening to hyperspectral pansharpening, to the sharpening of thermal images through an optical image, and to the fusion of optical and microwave image data from SAR are discussed.

4. **Image Registration and Interpolation.**
The basic problems occurring when two images are coregistered are concisely reviewed together with the related solutions. The theoretical fundamentals of digital interpolation, aimed at making the images to be merged, perfectly superimposed by pixel, are briefly recalled. After that, a survey of the most widely used interpolation kernels for resampling of the lower-spatial resolution dataset to the scale of the higher-resolution dataset is performed. Eventually, an analysis of the impact of the type of interpolation and of possible misregistration, either inherent in the datasets or induced by inadequate interpolation, on the quality of fusion is presented.

5. **Multiresolution Analysis for Image Fusion.**
The theoretical fundamentals and the definition of multiresolution analysis is presented. Focusing on wavelet analysis, which specifically is a type of multiresolution analysis, but in practice is a synonym of the relationship between wavelet analysis and filter banks, is highlighted. The terms *decimated* and *undecimated* analyses are explained. The differences between *orthogonal* and *biorthogonal* transforms are pointed out. The *à-trous* wavelet (ATW) analysis and the related transform are introduced and discussed. Nonseparable wavelet transforms (e.g., curvelet, contourlet) are briefly addressed. Eventually, Gaussian and Laplacian pyramids, not only for dyadic (octave) spatial analysis, but also generalized to a *fractional* analysis, that is, to an analysis with fractional ratios between adjacent scales, are reviewed and their suitability for image fusion is highlighted. Apart from the nonseparable case, four types of analysis have demonstrated their usefulness for RS image fusion: discrete wavelet transform (DWT), undecimated discrete wavelet transform (UDWT), ATW, and generalized Laplacian pyramid (GLP).

6. **Spectral Transforms for Multiband Image Fusion.**
   The fundamentals of multispectral image fusion based on spectral transforms are recalled. Transformations of three-band spectral pixels, that is, color pixels, are defined from the red-green-blue (RGB) representation of colors. The intensity-hue-saturation (IHS) transform and its nonlinear version, the brightness-hue-saturation (BHS) transform, can be defined and implemented in several ways. As main contribution of this chapter, multispectral fusion based on IHS transformations, either linear or not, may be extended to an arbitrary number of spectral bands, by exploiting the property that only the intensity component I is affected by fusion. This allows users to define a *generalized* IHS, which is also a *fast* IHS, because the spectral transform, which strictly does not exist if there are more than three bands, is never calculated, either back or forth, to yield fusion products. The generalized IHS transform features as degrees of freedom the set of spectral weights that yield I from the MS channels. Such weights may be optimized, for example, to attain the best spectral matching between the panchromatic image P and the intensity I. Principal component analysis (PCA), which holds for any number of bands, is reviewed and its decorrelation properties are highlighted, together with its use for multispectral fusion. Ultimately, the Gram-Schmidt orthogonalization procedure is found to be the basis of a spectral transform that has led to a very popular fusion method: Gram-Schmidt spectral sharpening, still suitable for an arbitrary number of bands.

7. **Pansharpening of Multispectral Images.**
   The integration of a low-resolution multispectral (MS) image and a high-resolution panchromatic image (Pan) is addressed. Classical approaches in the literature are critically reviewed. Existing categorizations of pansharpening methods are encompassed by a new framework recently proposed by the authors. The features of methods and related data products depend on the way spatial details to be injected in the MS bands are extracted from Pan. Methods of "type 1" use digital spatial filtering of Pan and roughly correspond to the class of multiresolution analysis (MRA) or ARSIS methods. Though all labeled as wavelet transform, DWT, UDWT, and ATW, or GLP require a unique lowpass filter, besides the interpolation filter, in most cases of practical interest. Methods of "type 2" do not use spatial filters or MRA but exploit a spectral transformation (IHS, PCA, Gram-Schmidt, etc.) to produce an intensity component I from the MS bands, such that details are given by $P - I$, without spatial filtering of the Pan. Optimization strategies for methods of type 1 and type 2 are presented and discussed. In the former case, key point is the use of lowpass filters matching the modulation transfer functions (MTF) of the MS channels. In the latter case, a winning strategy is the design of an intensity component matching the statistics of the lowpass-filtered Pan, in order to achieve spectral matching. Both types

of methods benefit from proper detail injection gains achieved through suitable modeling of the relationships between MS and Pan. The majority of existing pansharpening methods, including hybrid methods (e.g., IHS + wavelet, PCA + wavelet, etc.) can be accommodated in either of the classes, though in most cases they belong to MRA, and are optimized by following standard guidelines.

8. **Pansharpening of Hyperspectral Images.**
The integration of a low-resolution hyperspectral (HS) image and a high-resolution panchromatic image is addressed. The lack of physical congruence, that is, of spectral overlap, between the short wave infrared (SWIR) channels of HS and Pan, makes the most of the otherwise valuable pansharpening methods to be definitely unsuitable for this task. Component substitution (CS) methods are less powerful than MRA methods, mainly because spectral mismatch between I and Pan may occur. To avoid this, an extension of the generalized IHS fusion method, which features as many intensity components as the spectral channels, seems to be little penalized by the spectral mismatch, though it requires joint optimization of a number of parameters equal to the squared number of spectral bands, which is extremely large for HS images. Conversely, MRA methods equipped with an injection model capable of constraining the spectral angle between resampled original and fused pixel vectors to an identically zero value, are a promising solution that is investigated. Concerning quality measurements, spectral consistency is more crucial than spatial consistency.

9. **Effects of Aliasing and Misalignments on Pansharpening.**
As a result of original studies recently carried out by the authors, it can be proven that all CS methods are little sensitive to aliasing of MS data and misregistration (intrinsic or induced by improper interpolation) between the MS and Pan images. Conversely, MRA methods are sensitive to misregistration and also to aliasing, unless the GLP is employed for MRA. The decimation stage of GLP can be designed in such a way that aliasing patterns of opposite sign to those of MS are generated in the spatial details of Pan that will be injected into MS. Intrinsic imperfections of the datasets, like aliasing and coregistration errors, explain the popularity among users of CS methods with respect to MRA methods. On the other side, unlike all MRA methods, CS methods are highly sensitive to *temporal* misalignments, that is, to MS and Pan acquired not at the same time. With CS methods, colors of the pansharpened MS image will be intermediate between the colors of the original MS image at time 1 and the colors of the MS image at time 2, when the Pan image was acquired.

10. **Fusion of Images from Heterogeneous Sensors.**
The fusion of image datasets coming from instruments that do not share

wavelengths intervals and/or imaging mechanism is an extremely challenging section of RS image fusion. The degree of heterogeneity between instruments rules the progressive abandon of pansharpening-like methods that exploit homogeneous datasets. Two cases of interest are recognized as:

(a) integration of thermal (monoband or multiband) imagery and optical images, either visible/near-infrared channels or a broadband panchromatic image. General methodologies are borrowed from pansharpening, but physical congruence must be restated depending on application issues in practical case studies. Quality assessment is feasible because the fusion product is expected to match a thermal image acquired from a lower height by the same instrument. An example of fusion of V-NIR and multiband thermal images acquired from the ASTER satellite is presented and discussed.

(b) integration of optical (V-NIR and possibly Pan) and SAR images is addressed. Depending on the relative spatial resolutions of the datasets and of the application goal of fusion, different strategies, not necessarily borrowed from pansharpening, can be adopted. In both cases, possibilities and limitations of applying the resulting fused data to visual analysis and classification are investigated. The most investigated problem is how to produce an optical image enhanced by some feature extracted from the SAR image. However, the dual problem, that is, producing a SAR image spectrally and/or radiometrically enhanced by exploiting the optical image, is expected to catalyze the attention of researchers over the coming years. In both cases, quality assessment is complicated by the fact that there is no instrument that collects an image similar to the fusion product.

11. **New Trends of Remote Sensing Image Fusion.**
New directions in image fusion are addressed. Up-to-date methods based on sparse signal representations, compressed sensing, Bayesian approaches, and new spectral transformations are reviewed and compared to classical methods in terms of performance and computational complexity. A critical analysis is presented on modeling errors due to simplifications that are intentionally introduced to reduce the computational complexity, in particular on Bayesian and variational methods. An investigation on the unconstrained/constrained optimization methodologies driving the fusion process is presented and the impact of the underlying assumptions for RS datasets and applications is discussed.

Problems due to iterative processing and numerical instability are analyzed in the framework of the super-resolution fusion paradigm. In this context, image fusion is generally a severely ill-posed problem because of the insufficient number of low-resolution images, ill-conditioned registration, and unknown observational model parameters. Various regularization methods are introduced to further stabilize the inversion of this ill-posed problem. Alternative pansharpening methods adopt new spectral representations of multiband images, which assume that an image can be modeled as a linear composite of a foreground component carrying the spatial detail and a background color component conveying the sensed spectral diversity. The advantage of this new approach is that it does not rely on any image formation assumption as in the case of restoration or variational or sparse decomposition approaches.

12. **Conclusions and Perspectives.**
The main conclusions that can be drawn after reading this volume can be summarized in the following key points:

- Fusion methods have been motivated by progresses in sensor technologies. Early methods represent a first generation and are unconstrained from objective quality measurements.

- Definitions of quality of fusion products are highly crucial and should rely on widely agreed concepts and on consistency measurements.

- The majority of fusion methods, including hybrid methods, can be accommodated in a unique framework that may be parametrically optimized, and constitutes the second generation of fusion methods.

- An established theory and practice of pansharpening, together with availability of data from upcoming missions, shall open wider horizons to HS pansharpening, thermal and optical image fusion, and especially to SAR and optical image fusion.

- New paradigms of signal / image processing, like compressed sensing, are largely investigated nowadays and it is expected that in some years they will provide a third generation of methods, more powerful than those of the second generation, but computationally massive and presently unaffordable.

The authors believe that the last item will mark the progress in RS image fusion in the coming years.

1.4   Concluding Remarks

After reading this brief chapter of *Instructions for Use*, the presentation of materials and methods in the following chapters should be expedited and finalized to the goal of understanding why some choices were undertaken in the past and how better strategies can be devised today.

Ultimately, we wish to remind the reader that the present volume covers the birth and development of three generations of RS image fusion. Established textbooks are mainly concerned with the earliest generation of methods. The second generation is the objective of this book, while the first generation is reported concisely, mainly for historical completeness and tutorial purposes. The third generation is still to come. Chapter 11 is devoted to such new trends of RS image fusion, which are still relatively far from reaching the status of a third generation of methods.

Chapter 2

Sensors and Image Data Products

2.1   Introduction

The objective of this chapter is to provide an overview of the basic aspects involved in the acquisition and in the production of remote sensing (RS) data. In this context the word *sensor* will be used as a synonym of acquisition system. The basic principles of RS and a definition of the physical entities involved in the acquisition process will be briefly recalled, by focusing on the main parameters of sensors, which determine the characteristics of the acquired images. Some aspects related to sensors will be explained, in order to understand the properties of the data that are encountered in RS applications. The goal is introducing the reader to basic concepts that are relevant for RS image fusion. Optical sensors are first presented by making a distinction between reflected radiance and thermal imaging sensors. Afterward, such active sensors as synthetic aperture radar (SAR) and LiDAR are dealt with. Although LiDAR is an optical instrument, its importance lies in its feature of being *active*. When the various sensor typologies are presented, a brief review will be done of some imaging sensors that are relevant for *pansharpening* fusion. Their main characteristics are presented in order to delineate the framework of the past established work in image fusion and motivate the present and future research activity.

2.2   Basic Concepts

The basic task of RS is to measure and record the energy reflected or emitted by the Earth's surface by means of sensors hosted on airplanes or orbiting on satellite platforms. The energy at the sensor is the result of the interaction of several physical aspects, which are mainly ground properties, scene illumination, relief, and atmospheric influence. Each measurement is usually associated to a coordinate system thus originating a function $f(x, y)$ representing an image. In general, f depends not only on the spatial coordinates (x, y) but also on the wavelength (λ), or wavelength interval for incoherent systems, in which the sensor operates. In general, a sensor can acquire data not only in the range between two given wavelengths (spectral band or channel) but in several channels simultaneously as in the case of multispectral (MS) or hyperspectral (HS) instruments. Whenever multitemporal observations are considered, f will depend also on time (t). In instruments providing *range* information, in some cases, f does not represent an energy, but, for example, the time difference between two energy pulses, as in the case of an altimeter, which derives height from the time of flight of the returning signal. For our

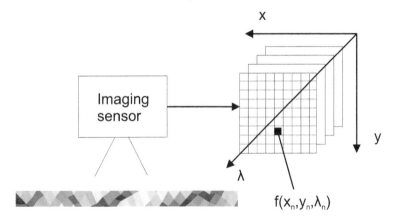

FIGURE 2.1: Basic imaging sensor scheme. The upcoming radiation is imaged at the detector and converted to digital values.

purposes, unless explicitly indicated, f will be representative of an energy, whose dependency will be on the independent variables (x, y, λ).

Modern imaging sensors record data in a digital format. Spatial coordinates (x, y) are thus sampled to discrete values and f is also quantized to discrete values. Sampling theory [197, 230] and quantization theory [176, 182] represent the reference frameworks in which these operations are formally studied and defined. Sampling theory establishes the ideal conditions by which an analog signal can be sampled and successively reconstructed without introducing any distortion. When such ideal conditions are not met because of an insufficient sampling rate some distortions occur. Dually, quantization theory determines the distortions derived from quantizing the continuous values associated to the discrete spatial coordinates. Unlike sampling that could ideally be a reversible process, quantization always introduces some distortions. Basic sampling and quantization aspects will be briefly considered in Sections 2.2.2 and 2.2.3.

Also wavelength (λ) will be assumed as a discrete variable whose discretization will be addressed in Section 2.2.4. As a direct consequence of sampling and quantization, image data are made up of discrete picture elements, or *pixels*, each of which is characterized by a discrete value, as shown in Figure 2.1. From the user point of view, the main parameters of a remote sensing acquisition system are the number of bands available, the spectral range of each band usually defined by the wavelength interval, the spatial resolution (assumed in practice as the pixel size even if not correctly as discussed in Section 2.2.2) and the radiometric resolution. In such application as change detection and monitoring also the repeatability of the observations (temporal resolution or revisiting time) is important. All these aspects are addressed in the next sections.

2.2.1   Definition of Terms

The terms regarding electromagnetic radiation (EMR) one may encounter in remote sensing are often utilized in an intuitive way. Some of the terms that occur more frequently are reported in the following. For an in-depth explanation the reader can refer to the literature [63, 180, 181, 219, 236, 238].

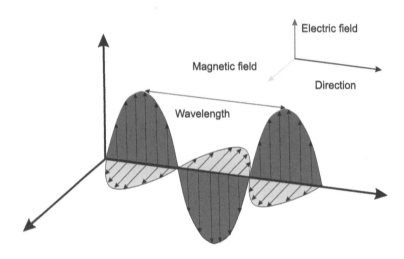

FIGURE 2.2: Electromagnetic radiation. Electric and magnetic field oscillate with a phase perpendicular to one another in a plane perpendicular to the direction of propagation.

EMR propagates as a transverse wave characterized by two components, the electric field and the magnetic field. These two fields oscillate in a phase perpendicular to one another and stay in a plane perpendicular to the direction of propagation of the EMR, as shown in Figure 2.2. Polarization is a property of an EMR and is conventionally defined by the direction of the electric field. The electric field may be oriented in a single direction (linear polarization), or it may rotate as the wave travels (circular or elliptical polarization). Just as an example, the polarization of the EMR reported in Figure 2.2 is vertical. Polarization is commonly used in microwave remote sensing to characterize the properties of backscattered radiation. The range of all possible frequencies of an EMR is denoted as the electromagnetic spectrum.

Radiant energy is the energy associated with EMR and is measured in Joules. The amount of radiant energy emitted, transmitted, or received per unit time is denoted as radiant flux and is measured in Watts.

The interaction between EMR and a surface can be described by means of *radiant flux density*. Radiant flux density is the magnitude of the radiant flux that is incident upon or, conversely, emitted by a surface of unit area. It is measured in Watts per square meter ($W \cdot m^{-2}$).

When radiant energy hits a surface, radiant flux density is denoted as *irradiance*. If the energy is emitted by the surface, as it occurs for thermal energy emitted by the Earth, or reflected by the Earth as it occurs for solar energy, then the term *radiant exitance* or *radiant emittance* is utilized.

A fundamental concept when dealing with sensors is *radiance*. Radiance identifies the radiant flux density reflected or emitted by the unit area of a surface (usually the Earth's surface) as viewed through a unit solid angle. The solid angle is measured in steradians. A steradian is defined as the solid angle subtended at the center of a unit sphere by a unit area on its surface. The radiance can be thus defined as the *radiant exitance* or *radiant emittance* for solid angle. Consequently, it is expressed in Watts per square meter per steradian ($W \cdot m^{-2} \cdot sr^{-1}$). If we assume for simplicity that radiance is due to the contribution of only one source (solar reflection or Earth emittance) it will depend on wavelength according to Planck's law. The radiance measured at the sensor will be thus integrated on the wavelength range the sensor is operating. In order to refer radiance to the unit of wavelength it is convenient to introduce the concept of *spectral radiance*. Spectral radiance is defined as the radiance per unit wavelength interval and is expressed in $W \cdot m^{-2} \cdot sr^{-1} \cdot \mu m^{-1}$. Spectral radiance is the entity that is usually delivered when data are disseminated by producers.

Analogous to radiance, all the entities that have been introduced can be defined per unit wavelength interval. In that case the attribute *spectral* is prefixed to the name of the entity and its unit of measure is modified accordingly.

Reflectance is defined as the ratio of the radiant emittance of an object and of its irradiance and is dimensionless. Unlike other entities that depend on environmental conditions, reflectance can be adopted in image analysis in order to eliminate the effects of variable irradiance. It is realized immediately that the reflectance is fundamental when comparing images observed at different times. In fact, observed radiances would be affected by such observation parameters as solar angle, observation angle, solar distance, local time, atmosphere, and so on. Only once corrected for such effects and converted to reflectance values, can the images be properly analyzed and compared.

2.2.2 Spatial Resolution

The spatial resolution of an imaging system is not an easy concept to define [181]. Several criteria have been identified to found its definition. The geometrical properties of the imaging system, the ability to distinguish between point targets, the ability to measure the periodicity of repetitive targets, and the ability to measure the spectral properties of small targets [250] have been used to base a definition of spatial resolution. We shall focus our discussion on geometric aspects and on system impulse response. For a more in-depth explanation on this topic the reader can refer to the Bibliography [48, 104, 234, 250].

Intuitively, spatial resolution refers to the capability of distinguishing spatial features in an image and can be expressed as the minimum distance by which two separate objects are perceived as disjoint. If we focus on optical imaging sensors, the spatial resolution is generally related to the instantaneous field of view (IFOV) of the instrument: the angular IFOV is the solid angle through which a detector is sensitive to the incoming radiation, that is,

$$\text{IFOV} = \frac{d}{f} \tag{2.1}$$

where IFOV is expressed in radians, d is the dimension of the detector element, f is the focal length of the instrument, and (2.1) is valid for $d \ll f$. The segment length on the Earth's surface obtained by multiplying the IFOV by the distance of the instrument from the ground is denoted as spatial resolution or spatial IFOV as outlined in Figure 2.3(a).

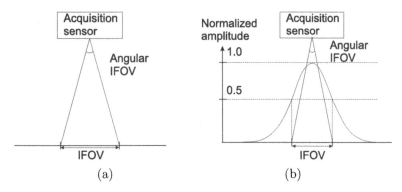

FIGURE 2.3: Geometric definition of IFOV (a) and by mean of PSF (b). In (b) IFOV is the full width of the PSF at half its maximum.

While IFOV refers to the elementary resolution cell, it is important to also define the field of view (FOV) of an instrument. The angular FOV is the total angle by which a sensor can observe a scene. The FOV is partitioned into the set of all the IFOVs. For push broom sensors (see Section 2.3.2) the FOV is practically the number of detector elements multiplied by the IFOV. The spatial FOV can be defined accordingly. For push broom sensors, the spatial FOV in the across-track direction is denoted as swath width or simply *swath*.

It should be noted that the spatial IFOV depends on the height of the platform and on the acquisition geometry: elements of the scene located at different positions of the FOV exhibit a different spatial resolution. This last aspect is usually crucial for airborne platforms (large FOV) and negligible for satellite sensors (small FOV).

Although the angular IFOV depends mainly on the instrument and is independent of the altitude of the platform and thus more appropriate than spatial IFOV to define the resolution, spatial IFOV is usually adopted in practice to identify the spatial resolution of a sensor. This occurs especially for sensors orbiting on satellites since we can assume that the altitude is nearly steady, the FOV is small, and thus the spatial resolution practically constant for a given platform.

As evidenced in Section 2.3.2, the spatial resolution along the flight direction is usually different from the resolution across the flight direction. As a consequence it is useful to consider two IFOVs, one along and the other across the flight direction, respectively.

It should be noted that although usually related to it, pixel size should not be confused with spatial resolution. In practice, the term *pixel* refers to the final output image of the acquisition system on which geometric corrections and resampling are usually performed before its delivery to users.

Another concept that is often used in remote sensing is *scale*. It expresses the ratio between a geometric representation of the reality (typically a map) and the reality itself. A map with a scale 1:1000 means that, once a geometric feature is measured on the map with any unit of measure, its real dimensions can be obtained by multiplying the result by 1000. Although scale can be related to resolution, the two concepts must not be confused. For example, an image with a given resolution can be represented at any scale. Of course, the image will appear smooth when spatial details are not sharp enough at that scale, that is, the spatial resolution is too coarse at that scale. When producing a map from an image of a given resolution, a scale exists, over which the resolution will no longer be adequate to represent reality with negligible geometric errors. This limit is indicated in Table 2.1 for some typical optical sensors. Obviously, a small scale denotes a coarse representation and vice-versa.

The definition of spatial resolution through the IFOV is based on geometric considerations. It is rather immediate and intuitive but does not consider all the aspects of the acquisition system. A further insight is considered in Section 2.2.2.1 by means of the characterization of the imaging system in the space and spatial frequency domains. Their importance goes beyond resolution aspects since they have a direct impact on the design of the image processing algorithms; concerning pansharpening schemes, especially on those that are based on multiresolution analysis (MRA).

2.2.2.1 Point spread function

Let us consider the acquisition process by which an input scene is sensed by the detector located in the image focal plane of the sensor. Each object of the scene is projected on the detector with a scaling factor and a blurring effect that is characteristic of the system optics. In order to characterize the system from a mathematical point of view, two main hypotheses are assumed. The first is that the system is *linear*, that is, the superposition theorem holds. If we consider the process of imaging several sources, the image that is produced when considering the contribution of all the sources together is the same of that produced when each source is imaged separately and all the images are then added each other. The second assumption is that the system is *shift invariant*, which means the response of the system is the same when considering different regions of the image focal plane. Under these assumptions, the behavior of the system is characterized by its characteristic blurring function, which is the system's response to an input bright point and is thus denoted as a *point spread function* (PSF) or *system impulse response*. The linearity hypothesis can be assumed for most of the imaging systems involved in the acquisition of remote sensing data, while the shift invariance hypothesis is practically reasonable for most imaging systems once diffraction effects caused by aperture, spatial distortions due to lens geometry, and dispersion effects caused by the dependency of light by its wavelength can be considered as negligible.

The PSF completely characterizes the optical properties of the acquisition system. If we consider Figure 2.3(b) where a normalized PSF is reported for a 1-dimensional case, the spatial resolution can be defined as the full width of the PSF at half its maximum. This is the definition of the IFOV of the optical system by means of the PSF. Such a definition is more significant than (2.1) that is mainly based on geometric considerations. It is related to the whole optical system and not only on the physical size of the detector elements. In order to point out possible differences between the two definitions, let us consider the case in which the IFOV defined by the PSF is significantly larger than the geometric IFOV. In this situation, the resolution of the system would not be consistent with the dimension of the detector element, which should be larger. Such a system would uselessly oversample the input scene with an unjustified cost in terms of transmission and storage. Conversely, when the IFOV defined by the PSF is significantly narrower than the geometric IFOV, the dimension of the detector element should be smaller. Otherwise, the acquisition system would undersample the imaged scene and cause a distortion effect known as *aliasing*. When the system is properly designed, the PSF IFOV and the geometric IFOV are consistent in value.

2.2.2.2   Modulation transfer function

Let us now consider the modulation transfer function (MTF) of an optical system. The MTF is equal to the the modulus of the optical transfer function (OTF) that is defined as the Fourier transform of the PSF of the imaging system (see Figure 2.4 for two examples of MTF that are discussed below). In most practical cases, that is, when the PSF is symmetric, the OTF is real and equal to the MTF. The MTF completely characterizes the optical system in the Fourier domain in the same way that the PSF does in the spatial domain. The advantage of working in the Fourier domain is that most of the operations necessary to model the entire system, which are convolutions in the spatial domain, become multiplications in the Fourier domain. This is a direct consequence of the linearity and shift invariance assumption.

It is important to underline that the scene to be imaged at the sensor can be modeled as the product of its Fourier transform with the system OTF (in

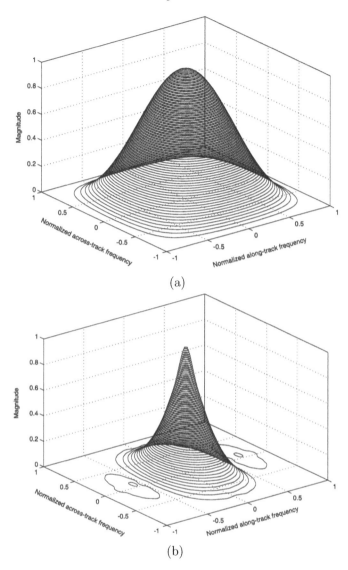

FIGURE 2.4: Example of ideal and real sensor MTFs. Ideal (isotropic) with magnitude equal to 0.5 at cutoff Nyquist frequency (a); typical (anisotropic) with magnitude 0.2 at cutoff Nyquist frequency (across-track) (b). All frequency scales are normalized to sampling frequency (twice the Nyquist frequency).

practice the MTF) in the same way it can be modeled by convolution with the PSF in the spatial domain. Since the resulting signal is to be sampled according to sampling theory by limiting as much as possible the number of samples, the MTF properties and the choice of the sampling frequency (also denoted as the Nyquist rate) will define the characteristics of the whole system. In particular, if the Nyquist rate is too low, a significant aliasing effect caused by an insufficient sampling because of the overlapping replicas of the Fourier spectra will result.

Just as examples, we report in Figure 2.4 the plots of two MTFs. Let us assume that the Fourier spectrum of the radiance of the imaged scene at the sensor is flat and normalized to 1 in order to simplify our considerations. Figure 2.4(a) shows a quasi-ideal MTF. For this MTF, the spectral replicas, which are originated by 2-D sampling of the radiance signal with sampling frequency along- and across-track equal to the Nyquist rate, should cross each other with magnitude values equal to 0.5 at the Nyquist frequency (half of the Nyquist rate). However, the scarce selectivity of the response of real systems prevents from using a Nyquist frequency with magnitude equal to 0.5 because of the aliasing phenomenon that would be generated. As a trade-off between maximum spatial resolution and minimum aliasing of the sampled signal, the Nyquist frequency is usually chosen such that the corresponding magnitude value is around 0.2. This situation is depicted in Figure 2.4(b) portraying the true MTF of an MS channel.

In the case of a system constituted by optics, a detector array, an electronic subsystem, the system MTF can be defined in the Fourier domain as:

$$MTF_{Sys} = MTF_{Opt} \cdot MTF_{Det} \cdot MTF_{Ele}$$

Each subsystem can be further subdivided into its components. For example, MTF_{Opt} can be expressed as the product of each optical element in order to evaluate its specific contribution. Depending on the depth of the analysis of the system, additional effects can be taken into account by introducing the corresponding MTF terms. For instance, in the case of push broom sensors, an MTF term is to be introduced to take into account of the motion of the platform(MTF_{Mot}) during the acquisition time. As a general consideration, the product of each MTF term causes a decrease of the overall system MTF. When sampling the input scene at the Nyquist rate, a value of the MTF is determined in correspondence of the Nyquist frequency. If this value is too high (> 0.4), aliasing may occur. As often happens when high-spatial resolution is required, this value could be too low (< 0.2). In this case, the acquired images could lack details and appear overly smooth.

Once the system MTF has been modeled with all the proper terms, the system PSF can be derived by taking its reverse Fourier transform.

An in-depth explanation on system MTF is beyond the scope of this book. Further references can be found in Slater [236] and Holst [135]. We highlight here that the knowledge of system MTF is a fundamental parameter for pan-sharpening algorithms. In fact, when algorithms take into account the MTF, the quality of the fused images strongly increases as discussed in Chapter 7.

2.2.2.3 Classification of sensors on spatial resolution

Eventually, we report in Table 2.1 a classification of sensors on the basis of their spatial resolution [189]. Table 2.1 is devoted to optical sensors but the classification can also be properly extended to cope with different acquisition systems such as a SAR or LiDAR scanner. The classification is organized into seven classes from extremely high resolution (EHR) to extremely low resolution (ELR). The resolution ranges are expressed in meters. The maximum significant scale of maps that can be derived from the acquired images is also reported together with examples of objects that can be distinguished. The last column of the table indicates some of the sensors that operate for the corresponding class of resolution.

As a final consideration, we note that a high-spatial resolution facilitates interpretation by a human observer while a very high resolution may lead to a high object diversity, which could cause problems when automated classification algorithms are applied to the data.

TABLE 2.1:   Classification of satellite sensors on spatial resolution. EHR: extremely high resolution; VHR: very high resolution; HR: high resolution; MeR: medium resolution; MoR: moderate resolution; VLR: very low resolution; ELR: extremely low resolution. Spatial resolution (Res.) is expressed in meters.

Res.	Type	Scale	Urban Object	RS Sensors
0.1 0.5	EHR	1:500 1:5.000	individuals, tiles, manhole duct cover	airborne
0.5 1.0	VHR	1:5.000 1:10.000	street lines, car, garage, small building, bush	Pléiades, GeoEye, IKONOS, QuickBird Worldview
1 4	HR	1:10.000 1:15.000	tree, building, truck, bus	Quickbird, Pléiades, GeoEye, IKONOS WorldView
4 12	MeR	1:15.000 1:25.000	complex, large building, industry, commercial	Rapideye, IRS, SPOT 5
12 50	MoR	1:25.000 1:100.000	vegetation cover, urban structure	Landsat TM, ETM+, ASTER
50 250	VLR	1:100.000 1:500.000	urbanized areas regional level	Landsat MSS, MODIS
>250	ELR	< 1:500.000	urban national level	NOAA AVHRR, Meteosat, MODIS

2.2.3   Radiometric Resolution

Radiometric resolution practically indicates the number of discernible discrete energy values and is also denoted as dynamic range. The finer the radiometric resolution of a sensor, the more sensitive it is to detecting small differences in reflected or emitted energy. Radiometric resolution is usually related to the signal-to-noise ratio (SNR) of the acquisition system: the better the SNR, the higher the dynamic range. Apart from reducing noise effects, the best way to improve SNR is to increase the energy of the signal collected by the sensor. Such energy usually decreases when spatial and spectral resolution increases. System designers are constantly looking for new solutions to this problem. One of the most interesting solutions that has been proposed in the last decades is to adopt time delay and integration sensors (TDI). A TDI charge-coupled device (CCD) is a multiline line scan CCD used in acquisition devices that captures high-speed movement with high sensitivity, which is unobtainable using conventional CCD arrays or even single-line-scan devices. During TDI operation, the photo charges gathered by each detector element are seamlessly shifted pixel to pixel down the detector, parallel to the axis of flow. By synchronizing the photocharge shift rate with the velocity of the flowing cell, the effect is similar to physically panning a camera. The advantage is twofold. On the one hand, the integration for several cycles (1 to 128 cycles are commonly available) increases the collected energy and thus SNR. On the other hand, the integration over several cycles, that is, on many detector elements, strongly reduces the annoying striping effects that commonly affects push broom sensors by averaging the unwanted varying gains of sensors.

In practice, the radiometric resolution is expressed by the number of bits, necessary to represent the range of available brightness values. Due to hardware constraints, and specifically of the numerical conversion process, the dynamic range is usually a power of 2. If the analog-to-digital converter (ADC) is realized with 12 bits, the corresponding dynamic range will be 2^{12} and thus 4096 levels. When the ADC is correctly dimensioned, the additive component of the noise should affect only the least significant bit (LSB) of the digital output signal.

The first satellite missions hosting optical sensors usually adopted 6 or 8 bit ADC. The ongoing missions operate with a radiometric resolution of 11 or 12 bits. The ADC output is usually denoted as digital count, since it represents just a number. Once the data have been properly calibrated, by considering sensor gains and dark currents, a radiance value is associated to them.

2.2.4 Spectral Resolution

Spectral resolution is the capability of a sensor of responding to a specific wavelength interval. Features of the ground can be identified by their reflected or emitted energy if the sensor is sensitive in the wavelength's range of the upcoming radiation. Depending on their spectral resolution, sensors can be classified as panchromatic, multispectral, superspectral, hyperspectral, and ultraspectral.

A panchromatic (Pan) sensor operates in a unique broadband. The most classic example is constituted by a sensor that integrates the radiation in the visible and near-infrared (V-NIR) spectral range. When such a sensor is realized by means of a CCD detector, the wavelength interval is typically [400 − 1000] nm. The wavelength range is thus several hundredths of nanometers. The poor spectral resolution of a Pan sensor is usually compensated by its spatial resolution that can be very high, that is, of the order of decimeters. Several commercial satellite data are usually distributed with a spatial resolution of 50 cm. Future missions are expected to acquire and process data well under this limit.

A multispectral (MS) sensor is expected to operate in several wavelength ranges (bands). Typical MS sensors exhibit 3 bands in the visible range. Wavelength ranges are narrower than the Pan sensors and one could expect to have wavelength ranges of the order of 50 nm. When the number of bands exceeds 10, the term *superspectral* is sometimes used. Due to system constraints, MS sensors operate with a spatial resolution which is lower than Pan sensors. This circumstance motivates the development of pansharpening algorithms, as successively discussed in Chapter 7.

When the spectral resolution increases and becomes better than 10 nm, the sensors are denoted as hyperspectral (HS). An HS sensor can exhibit hundreds of bands and a spectrum can be considered for each pixel of the imaged scene. Such measured spectra can be further related to laboratory spectra obtained by high-resolution spectrometers, in order to find the correspondence between field and laboratory measurements.

In case the spectral resolution of the imaging sensors increases up to the order of 1 nm or less, the instruments are denoted as *ultraspectral*. This is the case for instruments that operate on the principle of *Fourier Transform InfraRed* (FTIR) spectroscopy. Such sensors are typically based on a Michelson or a Sagnac interferometer that produces interferograms whose Fourier transform is the wavelength spectrum of the incoming radiation.

Figure 2.5 reports on a classification of sensors depending on their spectral bandwidth. Pan, MS, SS, HS, and US denote panchromatic, multispectral, superspectral, hyperspectral, and ultraspectral sensors, respectively. The spectral bandwidth decreases from Pan to US sensors, while, correspondingly, the number of bands increases.

2.2.5 Temporal Resolution

Temporal resolution is a key factor to be taken into account for applications. It is indicated also as *revisit time*, that is, the time range between two observations of the same scene. The revisit time is due mainly to orbital parameters and to the pointing capabilities of the satellite platforms. Apart from geostationary satellites that always observe the same scene with a spatial resolution of the order of kilometers and transmit a new observation several times an hour, the revisit time is crucial for most applications and especially in the case of monitoring. Indeed, disaster monitoring would require very short revisit times that are not allowed with orbiting polar platforms. The only fea-

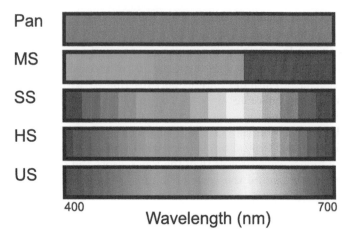

FIGURE 2.5: Sensor classification depends on the spectral bandwidth. Pan, MS, SS, HS, and US denote panchromatic, multispectral, superspectral, hyperspectral, and ultraspectral sensors, respectively.

sible way to increase the revisit rate is to consider a constellation of satellites with pointing capabilities.

Conversely from geostationary satellites, most satellites follow polar orbits, each having a specific repeat cycle and are Sun synchronous. Sun synchronous means that the local time a satellite crosses the equator is always the same for each orbit, that is, the satellite maintains the same relative position with the sun when crossing the equator. Such a choice is advantageous since the satellite passes in the same place at the same local hour (same local conditions facilitate such tasks as photo analysis) and furthermore, the satellite platform is in a stable position with respect to the Sun, thus facilitating some design aspects concerning attitude.

The temporal resolution of a polar orbiting satellite is determined by the joint choice of such orbital parameters as altitude, orbit shape, and inclination angle. Earth observing satellites usually have a near-circular polar orbit. The inclination angle between the orbital plane and the Earth equatorial plane is denoted as the inclination of the orbit and is the key parameter in determining the Earth coverage.

2.3   Acquisition Strategy

An important point concerns data acquisition. Aerial photography has been one of the first ways to obtain remote sensing data. Traditional air photographs were analogue in nature. They adopted a photographic film to record a range of light levels in terms of the differential response of silver halide particles in the film emulsion.

Nowadays, digital cameras are available with a spatial resolution and a dynamic range that is competitive or better than analogue photography. Of course, digital photography offers the advantage of an easy recording and flexible processing by means of a computer.

Modern remote sensing sensors are typically constituted by arrays (often matrices) of CCDs or solid state detector elements whose chemical composition is chosen on the basis of the requested wavelength response. Although they can operate on a camera-like principle, the most effective way they can acquire data is to exploit the motion of the platform on which they are mounted. Hereafter, the two basic operating modes, whisk broom and push broom, are examined.

2.3.1 Whisk Broom Sensors

Whisk broom sensors, also known as across-track scanners, measure the energy coming from the Earth's surface by scanning the ground in a series of lines constituted of elementary resolution cells as shown in Figure 2.6. The lines are perpendicular to the direction of the sensor platform (that is, across the swath). Each line is scanned from one side of the swath to the other, by utilizing such a device as a rotating mirror. As the platform moves forward over the ground, the succession of elementary resolution cells and swath lines forms the two-dimensional image of the Earth's surface. If a dispersion element is introduced in the optical system, the incoming reflected or emitted radiation can be separated into several spectral components that are detected independently. A bank or a linear array of detectors, each sensitive to a specific range of wavelengths, integrates the energy for each spectral band: for each resolution cell and for each band, the energy is converted in an electrical signal, digitized by an ADC, and recorded for subsequent processing. Several images can be produced (one for each spectral band) or an HS data cube (when the spectral sampling is narrow and typically of the order of 10 nm). The angular IFOV of the sensor and the altitude of the platform determine the spatial IFOV (the spatial resolution). The angular FOV is the sweep of the mirror. By taking into account the altitude of the platform, it determines the swath width of the instrument. Airborne scanners typically work with large angular FOV, while satellites need smaller FOV to sense large areas. Because of system constraints like platform speed, the dwell time (the time the energy from a ground resolution cell is integrated) is rather short and negatively influences the design of the spatial, spectral, and radiometric resolution of the instrument. One way to tackle this problem is to consider push broom sensors as discussed in Section 2.3.2.

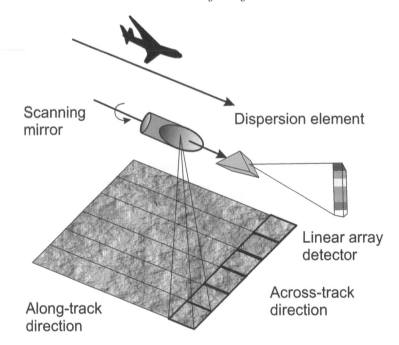

Scanning mirror

Dispersion element

Linear array detector

Across-track direction

Along-track direction

FIGURE 2.6: Scheme of a hyperspectral whisk broom sensor. The image is produced by the combined effect of the scanning mirror (across-track) and of the platform movement (along-track).

2.3.2   Push Broom Sensors

The basic idea of push broom sensors is to take advantage from the forward motion of the platform in the along-track direction to acquire and record successive scan lines of pixels and progressively form an image. They adopt a linear array of detectors located at the focal plane of the imaging system perpendicularly to the along-track direction. The extension of the array determines the angular FOV, and by taking into account the height of the platform, the swath width. The image is formed without resorting to any scan movement. The term push broom is due to the analogy of the motion of the detector array with a broom being pushed along a floor. Each detector element measures the energy of a single ground resolution cell and the angular IFOV determines the spatial resolution of the system. If a dispersion element is adopted to spectrally partition the incoming radiation, more linear arrays can be located at the focal plane thus obtaining an MS instrument with several spectral channels. For an HS instrument like that shown in Figure 2.7, a 2-D array (matrix) of sensors is utilized to record the energy of each pixel of a scan line (X dimension) for each wavelength (λ dimension). For a scan line, the energy integrated by each detector of the matrix is sampled electronically and digitally recorded. The sequence of the digitized data of all the scan lines (Y dimension correspond-ing to the platform direction) constitutes the hypercube of the acquired data. Push broom scanners exhibit several advantages with respect to whisk broom scanners. The array of detectors combined with the motion of the platform al-lows each detector to measure the energy from each ground resolution cell for a period of time (dwell time) longer than whisk broom sensors. More energy is integrated at the sensor and thus the radiometric resolution results improved. A greater energy also facilitates the achievement of smaller IFOVs and nar-rower bandwidths for each detector. Thus, finer spatial and spectral resolution can be achieved without heavily compromising radiometric resolution. Some advantages also occur because detector elements are usually solid-state micro-electronic devices. In fact, they are generally small, light, require low power, are more reliable, and last in time. Another important advantage of a push broom device is the absence of mechanical scanning, thus reducing vibrations during image acquisition and avoiding possible instabilities or damage to the moving parts. The price to pay for such advantages is the difficulty of calibra-tion. In fact, detector elements usually exhibit sensitivities that can vary by several percentages because of the manufacturing process. Cross-calibrating many thousands of detectors trying to achieve uniform sensitivity across the array is mandatory and complicated. In any case, a residual error in calibra-tion occurs and appears as striping effect on the acquired images. Indeed, destriping is usually necessary as one of the last steps of the calibration chain.

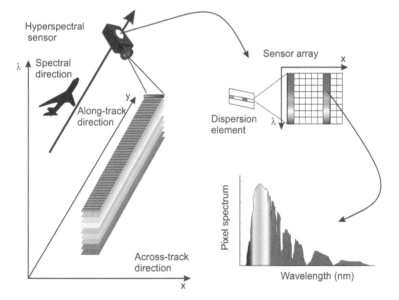

FIGURE 2.7: Scheme of a hyperspectral push broom sensor. The scanning mechanism in the across-track direction is no longer necessary since a whole row is acquired by the sensor array.

2.4   Optical Sensors

It is useful to distinguish three main modes by which the energy reaching a remote sensor is produced. In most scenarios it is the solar irradiation that is reflected by the Earth's surface. A second possibility concerns the energy that is emitted by any body because of its temperature. Eventually, the radiation at the sensor can be the consequence of a scattering mechanism produced at the Earth's surface by such an illumination source as a radar or laser.

When electromagnetic radiation hits the surface of a material, three types of interaction may occur: reflection, absorption, and transmission. If we neglect for the moment the scattering effect of the atmosphere, remote sensing is mainly concerned with the reflected part of this radiance that reaches the sensor. Reflected radiance is a signature of the material and, if it is measured over a range of wavelengths, we can obtain the wavelength spectrum of the material itself.

As a consequence of the absorption/transmission mechanism, any body increases its temperature until a balance is reached between the absorbed and transmitted radiation. Apart from conduction or convective effects, we are interested in the radiation of the body that is emitted as electromagnetic radiation since it can be measured by a remote sensor. This radiation has a power spectrum that depends on the temperature, according to Planck's law, and on the emissivity of the body. If we consider a *black body*, that is, an ideal body whose emissivity is 1 at all wavelengths, its power spectrum is fully characterized by Planck's law, that is, can be expressed as a function of λ and T so that:

$$I(\lambda, T) = \frac{2hc^2}{\lambda^5} \frac{1}{e^{\frac{hc}{\lambda k T}} - 1} \tag{2.2}$$

where I denotes the *spectral radiance*, that is, the energy per surface area per unit of wavelength per unit of time, T is the temperature of a black body expressed in Kelvin degree, h is the Planck's constant, c is the speed of light, k is the Boltzmann constant, and λ is the wavelength of the emitted radiation.

$I(\lambda, T)$ is usually plotted as a function of λ for a given T, as shown in Figure 2.8. The scales are logarithmic. The dashed diagonal line indicates the graphical representation of Wien's law, by which the wavelength at which the maximum of emission occurs increases along wavelength according to the law:

$$\lambda_{max} T = 2898$$

in which λ_{max} and T are expressed in μm and Kelvin degrees, respectively.

From this plot some fundamental considerations on remote sensing principles arise. The spectral emissivity of the Sun is similar to the curve of a black body at 5800 K. Because of solar gas absorption and terrestrial atmospheric

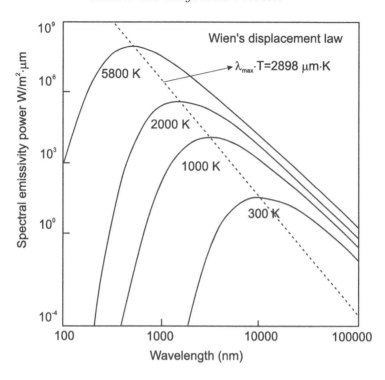

FIGURE 2.8: Spectral black body emissivity power. The maxima of emissivity follow Wien's law.

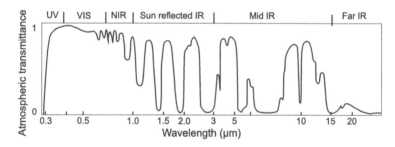

FIGURE 2.9: Atmospheric transmittance plotted from UV to far IR spectral regions.

transmittance, the shape of the spectrum is modified when reaching the Earth surface. Of course, its power is also scaled because of the Sun-Earth distance. Because of these effects and taking into account atmospheric transmittance (Figure 2.9), the result is that a sensor that observes the Earth surface can collect Earth reflected sun energy or Earth emitted thermal energy. Because of considerations on the power of this energy [236] the result is that a sensor

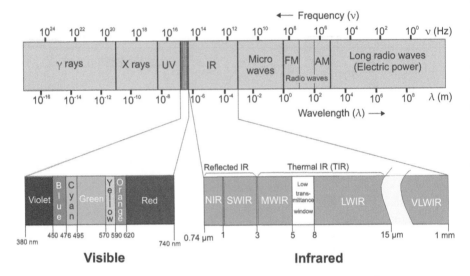

FIGURE 2.10: Electromagnetic spectrum. Visible and infrared spectral regions, in which most optical sensors work, are expanded.

practically collects reflected energy up to 3μm, mainly reflected energy with little influence of thermal energy in the range $[3-5]\mu$m and thermal energy only in the range $[8-12]\mu$m and beyond. The range $[5-8]\mu$m in which there is a balance between the reflected and emitted energy is practically useless because it does not correspond to a transmittance atmospheric window as can be observed in Figure 2.9.

By analyzing Planck's law, it can be observed that any body emits a radiation depending on its temperature. According to Wien's law, the wavelength for which the emission is maximum decreases for increasing temperatures and vice-versa, and the electromagnetic spectrum spans several order of magnitude in wavelength and correspondingly in frequency as shown in Figure 2.10. The spectral range in which remote sensing observations usually occur is expanded in Figure 2.10 between the wavelength of 380 nm and 1 mm. Apart from the visible spectral region that is characterized by colors as they are perceived by the human eye, some other spectral regions are evidenced. The reflected infrared (IR) region ($[0.74-3]\mu$m) is divided in the near IR (NIR) and short wave IR (SWIR) regions. Longer wavelengths ($[3-15]\mu$m) identify thermal IR (TIR) spectral region, which is partitioned in medium wave IR (MWIR) and long wave IR (LWIR) region. Among them a low transmittance window mainly due to water vapor absorption is present ($[5-8]\mu$m).

2.4.1 Reflected Radiance Sensors

Reflected radiance sensors measure the solar radiation that is reflected by the Earth's surface or scattered by the atmosphere within the FOV of the sensor. This radiation is the result of several interaction processes, as shown in Figure 2.11. When solar radiation reaches the atmosphere, it encounters gas molecules, suspended dust particles, and aerosols whose effect is the partial scattering of the incoming radiation in all directions. Most of the remaining light is transmitted up to the surface, apart from some spectral regions where atmospheric gases are responsible for absorbing the radiation at specific wavelengths. Because of these effects, the illumination at the surface is the superposition of filtered solar radiation transmitted directly to the ground and of diffused light scattered from all parts of the sky. At the ground, the radiation is partially reflected depending on the type of soil it may encounter, that is, rock surfaces, vegetation, or other materials. Each material absorbs a portion of the radiation. The absorption depends on the specific material and is a function of wavelength, thus determining a spectral signature in the radiation that is reflected toward the sensor. Ultimately, the reflected radiation is subject to further scattering and absorption processes since it crosses the atmosphere layers again before being detected and measured by the sensor.

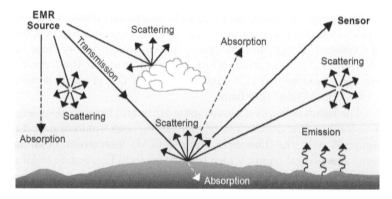

FIGURE 2.11: The radiance reaching a sensor is the result of complex interactions with the Earth's surface and the atmosphere.

2.4.1.1 Reflection

Reflection occurs when an EMR returns back after hitting a nontransparent surface. The nature of reflection depends on the surface characteristics and in particular on the size of surface irregularities (roughness or smoothness) with respect to the wavelength of the incident radiation. If the surface is smooth relative to wavelength, specular reflection occurs. Specular reflection redirects most of the incident radiation in a single direction according to the law by which the angle of incidence is equal to the angle of reflection. For visible radiation, specular reflection can occur with surfaces such as a mirror, smooth metal, or a calm water body [57]. When a surface is rough with respect to wavelength, a diffuse, or isotropic, reflection occurs and energy is scattered in all directions. For visible radiation, such natural surfaces as uniform grass or bare soil can behave as diffuse reflectors. A perfect diffused reflector is denoted as a *Lambertian* surface: the radiance is the same when observed from any angle, independent of the Sun elevation angle of the incident radiation. For such a surface the observed radiance, I, is proportional to the cosine of the elevation angle, θ, and to the radiance observed for $\theta = 0, I_0$, according to:

$$I = I_0 \cos \theta. \tag{2.3}$$

Although natural surfaces can be rarely considered as Lambertian, such an approximation is often adopted in remote sensing as a first-order approximation to compensate for the effects due to variations in the solar elevation angle. Such a correction is fundamental when comparing images of the same scene acquired at different local times, that is, with different elevation angles.

2.4.2   HR and VHR Sensors

Emphasis is given in this section to spatial resolution, rather than to spectral and temporal resolution, or better temporal coverage capability, of satellite systems. Hence, low/medium (spatial) resolution systems, like MODIS, MeRIS, SPOT-Vegetation, which are an heritage of the former millennium, are not mentioned in this context, while SPOT (1 to 5) is only briefly addressed throughout this book.

The Landsat family deserves some attention, because of its huge popularity over the last four decades. In particular, versions from Landsat 4 onward were equipped with the Thematic Mapper (TM) instrument. TM originally featured six spectral bands, which became seven in Landsat 5 (1=Blue, 2=Green, 3=Red, 4=NIR, 5=SWIR-1, 6=TIR, 7=SWIR-2). All bands have 30 m SSI, except Band 6 (120 m). Landsat 6 was the first Landsat satellite equipped with a Pan image. Because of its unsuccessful launch, researchers had to wait until 1999 for a Pan image at 15 m SSI, which was made available by Landsat 7 enhanced TM+ (ETM+). All data were packed in 8 bits. Nowadays, Landsat 7 is no longer active and Landsat 8 took its place in 2013. Despite technological progresses in sensors development, Landsat 8 has been chosen to be fully compatible with the Landsat generation, up to Landsat 7. In particular, Landsat 8 carries two instruments: the Operational Land Imager (OLI) sensor includes refined heritage bands, along with three new bands: a deep blue band for coastal/aerosol studies, a short wave infrared band for cirrus detection, and a quality assessment band. The thermal infrared sensor (TIRS) provides two thermal bands, resampled at 30 m for compatibility with the other bands, but originally collected at 100 m. The OLI panchromatic band is still geocoded at

15 m, but its bandwidth is narrower than that of Landsat 7 ETM+ (450 to 700 nm instead of 500 to 900 nm).

Technological evolution joined to some liberalization from the military to the commercial segment has allowed a number of American private-sector companies to launch and service their own satellites with a spatial resolution better than one meter (GeoEye, QuickBird, WorldView as examples). Further evolution carried to the launch of many other satellites with similar characteristics that can be denoted and labeled as very high-resolution sensors (VHR). Most of them are reported in Table 2.2 with their main characteristics. They usually host a single sensor that produces a panchromatic image and an MS image, typically with four spectral bands (more recently, 8 as in the case of WorldView). VHR satellites are designed with stereo capabilities in such a way they can acquire a stereo-pair images during the same orbit. The availability of such images has pushed digital photogrammetry and digital image processing to strictly cope together in order to improve the capability to generate DEMs and topographic mapping. Notwithstanding, designing a satellite mission takes several years, the evolution in this field is really quick and readers interested in up-to-date news should refer to such online information as *https://directory.eoportal.org/web/eoportal/satellite-missions/* or refer to the sites hosted by such spaces agencies as NASA, ESA, JAXA, DLR, CNES, and commercial companies involved in satellite missions.

In this section, some of the sensors reported in Table 2.2 are briefly reviewed. The focus is on those features that are relevant for the main target of RS image fusion, that is, for pansharpening.

Most of them adopt acquisition sensors featuring TDI. The TDI arrays prevent exposure saturation while maximizing the SNR over a wide range of angles and Earth albedos. Adopting the TDI concept provides a further advantage of reducing the striping effect that characterizes push broom sensors caused by detector nonuniform gains. Because of data averaging of the same picture element among several detectors, in the along-track direction, without introducing any loss in spatial resolution, gain differences are also averaged and thus reduced.

2.4.2.1 IKONOS

IKONOS was the first high-resolution satellite hosting an optical VHR sensor launched for commercial purposes. After the failure of the first launch, IKONOS-2 was successfully launched in 1999 and placed into orbit. IKONOS can acquire Pan images of 0.82 m resolution at nadir and MS images with four channels with a spatial resolution of 3.2 m at nadir. The system MTF of Pan sensor at Nyquist rate is 0.09. The data acquisitions of Pan and MS imagery data are *practically* simultaneous and fully coregistered. Indeed a time interval of some tenths of second occurs between the acquisition of the Pan and MS images. The consequence is a wake effect that appears on pansharpened images

TABLE 2.2:   Very high-spatial resolution satellite instruments. Spatial resolution is indicated at the nadir sub-satellite point. Revisiting time takes advantage of the pointing capabilities.

Satellite	Spat. Res. [m]	SSI [m]	Spectral Coverage [Km]	Swath Width [Km]	Rev. Time [day]	Notes
GeoEye-1	0.41	0.5	Panchromatic	15.2	3	Stereo
	1.65	2.0	Red, Green, Blue Near Infrared			
GeoEye-2	0.34		Panchromatic	14.2	3	Stereo
	1.36		Red, Green, Blue Near Infrared			
WorldView-1	0.50	0.50	Panchromatic	17.6	1.7	Stereo
WorldView-2	0.46	0.50	Panchromatic	16.4	1.1-3.7	Stereo
	1.8	2.0	Red, Green, Blue, Red Edge, Coastal, Yellow, NIR 1, 2			
WorldView-3	0.31	0.40	Panchromatic	13.1	1-4.5	Stereo
	1.24	1.60	Red, Green, Blue, Red Edge, Coastal, Yellow, NIR 1, 2			
	3.70	4.80	8 SWIR bands in the range 1195−2365 nm			
QuickBird	0.6	0.7	Panchromatic	16.5	1-3.5	Stereo
	2.4	2.8	Red, Green, Blue Near Infrared			
EROS-B		0.7	500-900 (pan)	7	3	Stereo
IKONOS	0.82	1.0	Panchromatic	11.3	1-3	Stereo
	3.2	4.0	Red, Green, Blue Near Infrared			
OrbView-3		1.0	Panchromatic	8	3	
		4.0	Red, Green, Blue Near Infrared			
KOMPSAT-2	1		Panchromatic	15	5	
	4		Red, Green, Blue Near Infrared			
Formosat-2	2		Panchromatic	24	1	
	8		Red, Green, Blue Near Infrared			
Cartosat-1	2.5		Panchromatic	30	5	Stereo
Pléiades-1A,B	0.7	0.5	Panchromatic	16.5	1-3.5	Stereo
	2.8	2.0	Red, Green, Blue Near Infrared			
RapidEye	6.5	5	Red, Green, Blue Red Edge, NIR			Stereo

in the presence of such moving objects as cars, trains, or airplanes that are observed at different positions in the Pan and MS data.

Images are distributed with an SSI of 1 m and 4 m, respectively, after ground image data processing that provides geocoding along with image

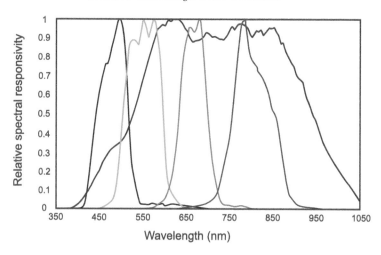

FIGURE 2.12: IKONOS's relative spectral responsivity. Pan's response is plotted as black. Color plots identify MS bands.

compensation for misregistration, image motion, radiometric correction, MTF compensation. The sensor can be tilted both along and across the track, and spatial resolutions < 1 m for the Pan image are possible for off-nadir pointing angles of less than 26 degrees. The relative spectral responsivity of the panchromatic and of the four MS channels are reported in Figure 2.12.

The satellite orbit is polar, near-circular, and Sun-synchronous at a nominal altitude of 681 km and with an inclination angle of 98.1 degrees. The satellite crosses the equator on a north-south at 10:30 local time. IKONOS data are quantized using with 11-bit and are distributed in a variety of forms ranging from standard system-corrected to geometrically corrected and stereo, for the production of DEM.

Due to high-spatial resolution and pointing capability, IKONOS images are suitable for small-area investigations, where they can effectively replace high-altitude air photography. Figure 2.12 reports the relative spectral responsivity of the IKONOS sensor. The Pan response is rather broad and is essentially due to the response of CCD detectors. Only a partial superposition of the blue band response occurs with the Pan response, thus possibly affecting the correlation existing between the blue and the Pan bands. This is also why the fused images produced by some of the forerunner pansharpening algorithms appears bluish on vegetated areas as it will be discussed in Chapter 7.

2.4.2.2 QuickBird

QuickBird was launched two years after IKONOS in October 2001. It is managed by DigitalGlobe company. As to IKONOS, it hosts a single instrument that acquires Pan images with a spatial resolution in the range 0.61 and

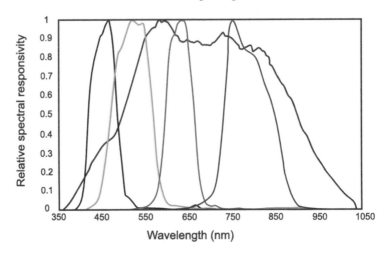

FIGURE 2.13: QuickBird's relative spectral responsivity. Pan's response is plotted as black. Color plots identify MS bands.

0.73 m and MS images with a spatial resolution in the range of 2.44 and 2.88 m, depending on the acquisition angle. Image data are quantized to 11 bits. The sensor has tilting capabilities along and across track, in order to allow the acquisition of stereo images and to guarantee a maximum revisit time of three and a half days. Imagery is available in basic (system corrected), standard (geometrically corrected to a map projection), and orthorectified forms. Figure 2.13 reports the relative spectral responsivity of QuickBird sensor. The two responses appear extremely similar. Only few minor differences occur because of band spectral wavelengths and extensions.

2.4.2.3   WorldView

WorldView-1 was launched in 2007 by DigitalGlobe Inc. It is equipped with a high-resolution (0.50 m) Pan sensor whose radiometric resolution is 11 bits. The limit of 0.50 m is just the best resolution allowed by U.S. government rules that prevent the sale to non-U.S. citizens of imagery with a resolution better than 50 cm. Figure 2.14 when compared with Figures 2.13 and 2.12 clearly shows that WorldView Pan is spectrally narrower than IKONOS and QuickBird. In particular, a nonnegligible range of wavelength has been filtered out beyond 900 nm.

In October 2009, DigitalGlobe Inc. launched WorldView-2. Conversely from WorldView-1, WorldView-2 operates both a Pan and an advanced MS sensor with eight bands (blue, green, red, NIR, red edge, coastal, yellow, and NIR-2). At nadir, the MS sensor has a spatial resolution of 1.8 m, while the Pan sensor has a resolution of 0.46 m. The radiometric resolution is 11 bits

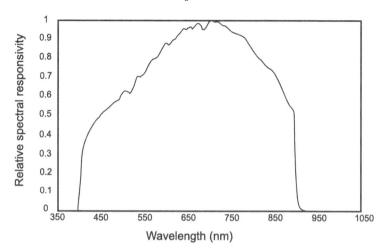

FIGURE 2.14: WorldView-1's relative spectral responsivity. WorldView-1 was equipped with the only Pan spectral channel.

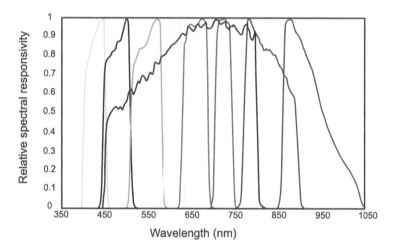

FIGURE 2.15: WorldView-2's relative spectral responsivity. Pan's response is plotted as black. Each MS band is plotted with a different color.

and has a swath width of 16.4 km. Figure 2.15 reports the relative spectral responsivity of WorldView-2.

The four primary MS bands include traditional blue, green, red, and near-infrared bands, which are similar but not identical to the QuickBird satellite. Four additional bands include a shorter wavelength blue band, centered at approximately 427 nm, called the coastal band for its applications in watercolor studies; a yellow band centered at approximately 608 nm; a red edge band

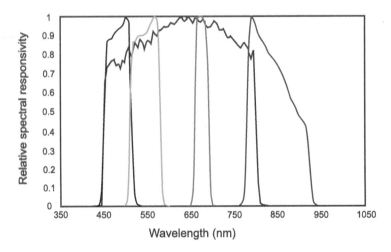

FIGURE 2.16: GeoEye's relative spectral responsivity. Pan's response is plotted as black. Each MS band is plotted with a different color.

strategically centered at approximately 724 nm at the onset of the high reflectivity portion of the vegetation response; and an additional near-infrared band, centered at approximately 949 nm, which is sensitive to atmospheric water vapor [254].

On August 13, 2014, WorldView-3 was successfully launched. WorldView-3 is a powered version of WorldView-2. In fact, its spatial resolution is enhanced since it is 0.31 m for the Pan sensor and 1.24 m for the MS sensor that operates in the VIS-NIR spectral range. The relative spectral responsivity of the two sensors is similar to that of WorldView-2 that is reported in Figure 2.15. As a further improvement, WorldView-3 imaging sensor was redesigned in order to add 8 spectral channels in the SWIR spectral region between 1195 and 2365 nm with a spatial resolution of 3.70 m. The joint exploitation of SWIR and VIS-NIR spectral channels provides WorldView-3 with notable capabilities in classification applications.

2.4.2.4   GeoEye

The GeoEye-1 satellite was launched on September 6, 2008. It provides panchromatic imagery with a resolution of 0.41 m and MS imagery with a resolution of 1.64 m, both at nadir. The radiometric resolution is 11 bits. Its relative spectral responsivity is reported in Figure 2.16. The Pan spectral width is similar to WorldView-2. The red channel is rather narrow when compared to other sensors. In particular, it is narrower than WorldView-2 narrow red channel designed to be better focused on the absorption of red light by chlorophyll in healthy plant materials.

2.4.2.5 Pléiades

The Pléiades program, set up by CNES, is the optic component of OR-FEO, a dual Earth observation system with metric resolution established as a cooperation program initiated between France and Italy. Pléiades comprises the constellation of two satellites (launched on 17/12/2011 and 2/12/2012, respectively) offering a spatial resolution at nadir of 0.7 m for the Pan sensor (2.8 m for MS channels) and a swath width of 20 km; the radiometric resolution is 12 bits. Their great agility of the system enables daily access all over the world, which is a critical need for defense and civil security applications, and a coverage capacity necessary for cartography applications at scales better than those accessible to SPOT family satellites. Pléiades relative spectral responsivity is reported in Figure 2.17. The Pan response is similar to GeoEye and WorldView-2. The blue and green channels are rather broad and exhibit a partial overlap that should increase the correlation of the resulting images. Also the red channel is far broader than those of GeoEye and WorldView-2.

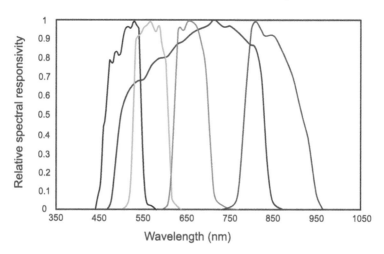

FIGURE 2.17: Pléiades's relative spectral responsivity. Pan's response is plotted as black. Color plots identify MS bands.

Pléiades satellites feature a stereoscopic acquisition capability to satisfy cartography needs and to allow the same information obtainable from aerial photography.

2.4.2.6 FormoSat-2 and KoMPSat

Other high-resolution satellite systems worth to be considered are FormoSat-2 and KoMPSat.

FormoSat-2 (formerly ROCSat-2, Republic of China Satellite-2) is an NSPO (National Space Program Office) of the Taiwan Earth imaging satellite with the objective to collect high-resolution panchromatic (2 m) and MS (8 m) imagery with a radiometric resolution of 12 bits for a great variety of applications such as in land use, agriculture and forestry, environmental monitoring, and natural disaster evaluation. The swath width is 24 Km. The satellite was launched in 2004 and is still operative.

The KoMPSat mission (Korean Multi-purpose Satellite) is constituted of two Earth observation satellites developed by the Korea Aerospace Research Institute (KARI). The first (KoMPSat-2) was launched in 2006 while the second (KoMPSat-3) on May 18, 2012. The characteristics of the two sensors are similar. Both have Pan and MS channels with similar spectral range. The spatial resolution of KoMPSat-3 is better than that of KoMPSat-2 (0.7 m vs. 1 m for the Pan channel and 2.8 m vs. 4 m for the MS bands).

2.4.3   Thermal Imaging Sensors

TIR imagers denote sensors collecting the radiation emitted from objects in the portion of the spectrum between about 3 to 14 μm. According to Planck's law (2.2), this is the typical spectral emission range corresponding to the temperature of objects located on the Earth's surface. Indeed, Planck's law refers to a black body, that is, an ideal body that absorbs all incident EMR, regardless of wavelength or incidence angle. If *reflectance* $\rho(\lambda)$ is defined as the fraction of incident electromagnetic power that is reflected by a surface at wavelength λ, then for a black body $\rho(\lambda) = 0$, $\forall \lambda$. Concerning emission, a black body is an ideal emitter; for a given temperature no body can emit more radiation than a black body. In addition, a black body is a diffuse emitter; the radiated energy is radiated isotropically at any direction. As a consequence of these definitions, if we define the emissivity as the ratio of the radiant energy emitted by a surface to that emitted by a black body at the same temperature, a black body has an emissivity $\varepsilon = 1$, while any body exhibits an emissivity $\varepsilon < 1$. If the emissivity of a surface is known, its temperature can be derived from (2.2). If the emissivity is not known, it can be assumed equal to 1 and in this case *brightness temperature* is derived, that is, the temperature the surface would exhibit if it were a black body.

Thermal sensors usually employ one or more internal temperature references (a quasi ideal black body) in order to compare the reference radiation with the measured radiation, in such a way measured radiation can be related to absolute radiant temperature. Concerning visual representation, an image of relative radiant temperatures is usually visualized in pseudo color or in black and white. Warmer temperatures are usually shown in light tones, and cooler temperatures in dark tones. However, notable exceptions exist: cold clouds are usually represented as white in order to make the work of photo analysts easier. Absolute temperature measurements may be calculated but require accurate calibration, measurement of the temperature of the reference bodies, knowledge of the thermal properties of the target, correction of geometric distortions and of radiometric effects.

Since the scattering of radiation depends on wavelengths, atmospheric scattering is minimal at the relatively long wavelength of thermal radiation when compared to visible radiation. In any case, the absorption of atmospheric gases forces thermal measurements to two spectral regions, 3 to 5 μm and 8 to 14 μm.

An undesired characteristic of TIR imagers is their poor spatial resolution. In fact, the energy collected at the sensor is low for thermal bands when compared to reflectance bands. In addition, thermal sensors are characterized by an additive noise that is not negligible. Such a noise can be reduced by cooling the sensors to low temperatures and by increasing the amount of energy coming from the surface. Unfortunately, this can be obtained at the price of increasing the IFOV and thus reducing the spatial resolution. In the presence of higher-resolution observations in different spectral bands, pansharpening can sometimes be adopted to increase the spatial resolution of the TIR image provided that a significant correlation exists between the two images as discussed in Chapter 10.

Unlike reflectance images, thermal images can be acquired during the day or night and can be used for such applications as military reconnaissance, forest fire mapping, and heat loss monitoring. Another important role that has been recognized for TIR images is the assessment of vegetation and the estimation of soil moisture by means of optical sensors.

2.5 Active Sensors

Whereas passive sensors operate by exploiting an external energy source (the Sun or the Earth's surface black body emission), active sensors provide by their own the energy, a fraction of which is returned back by the observed target and properly collected, processed, and recorded. Because of their importance, SAR systems and LiDAR scanners are considered in the following. SAR typically works in the microwave spectral region, which can be identified in Figure 2.10, through an antenna, while LiDAR operates by means of a laser in the optical frequency region.

2.5.1 Radar and Synthetic Aperture Radar

SAR (synthetic aperture radar) sensors operate on a different principle with respect to optical sensors and their complexity is significantly superior. SAR can provide high-resolution images independent of daylight, cloud coverage, and weather conditions. A comprehensive treatment of SAR theory is beyond the scope of this book. We report here only some basic concepts. The literature on SAR is really huge. Readers interested in more in-depth information on this topic can refer to the basic Bibliography and tutorials [77, 133, 193, 198, 201, 220, 247].

The forerunner of modern SAR was SLAR (side-looking airborne radar). Basically, SLAR is realized by means of ground-based radar.

SLAR operates by means of an antenna connected to a flying platform. The antenna is properly mounted in such a way that the energy radiates perpendicularly to the side of the platform, as shown in Figure 2.18, over a broad beam that defines the swath width of the recorded data. Conversely, the antenna beam is usually narrow in the direction parallel to the platform motion as it will determine resolution in the along-track direction. When the EMR arrives at the ground surface, part of it is backscattered to the platform where it is received by the same antenna and measured. The intensity of the signal is related to the properties of the surface and possible targets and is also influenced by such parameters of the transmitting/receiving system as frequency, polarization, and incidence angle.

Since there is no scanning mechanism, no spatial discrimination would exist in the across-track direction (range direction) unless the principle of radar is employed. Instead of transmitting a continuous signal, a short burst or pulses is transmitted at the frequency of the radar.

The pulses travel to the ground and are scattered back at the speed of light. The signal from the ground areas closer to the antenna are received earlier. When using pulses, the received signals are separable in time in such a way the strip of terrain across the swath is spatially resolved.

The transmitted pulses (ranging pulses) are repeated at a pulse repetition frequency (PRF). This is synchronized with the forward motion of the platform, so that contiguous strips of terrain across the swath are irradiated pulse by pulse. A constraint exists between the PRF and swath width in order to prevent signal ambiguity at the receiver: the higher the PRF the narrower the swath.

2.5.1.1 Across-track spatial resolution

The resolution across track (range resolution) is defined by the capability of distinguishing target pulses in the received backscattered signal. If two targets are too close, their backscattering will overlap and they will not be separable.

If we denote as Δr the separation in range of two targets, with Δt the separation in time between their echoes and with c the speed of light, then:

$$\Delta t = \frac{2\Delta r}{c}.$$

If τ is the pulse width, no separation is possible for $\Delta t > \tau$. By denoting the slant range resolution as r_r:

$$r_r = \frac{c\tau}{2}. \tag{2.4}$$

The slant range resolution is the resolution measured along the direction of

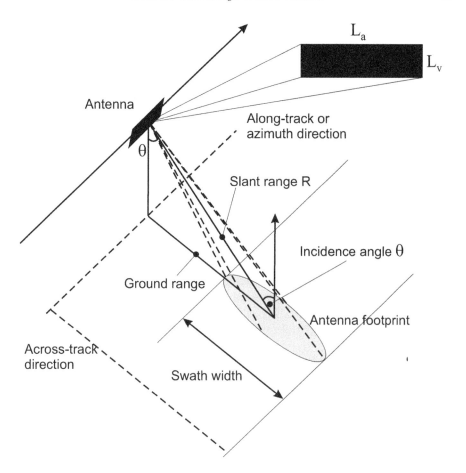

FIGURE 2.18: Geometry of a SAR system. Basic entities related to SAR's acquisition are identified in the figure.

propagation. From the user point of view, it is important to project this resolution on the ground. If θ is the incidence angle at the ground and the beam of radiation is incident locally, then the ground resolution in the range direction is:

$$r_g = \frac{c\tau}{2\sin\theta}.$$ (2.5)

Some important considerations are supported by (2.4) and (2.5). There is no spatial resolution if $\theta = 0$, that is, directly under the platform. The system works only with an incidence angle greater than 0 (SLAR mode) and there is usually a recommended value that is derived from the trade-off of system design. The slant and ground range resolutions are independent of the altitude of the platform but essentially depend on the characteristics of the transmitted signal. Ground range resolution is a function of the incidence angle and thus variable across the swath. Conversely from optical sensors, the resolution is better at far swaths (a greater θ determines a lower r_g).

A SLAR system can hardly work with interesting spatial resolutions because of energy and hardware constraints. The solution is to substitute the impulse signal with a chirp signal properly designed to introduce a modulation for the duration of the impulse τ. If matched filtering, based on correlating the returning signal with the chirp signal, is adopted at the receiver, it can be demonstrated that the resulting resolution becomes dependent on the bandwidth of the chirp signal. If B_c indicates the bandwidth of the chirp signal where $\frac{1}{B_c} \gg \tau$, (2.4) and (2.5) can be expressed as:

$$r_r = \frac{c}{2B_c} \tag{2.6}$$

and

$$r_g = \frac{c}{2B_c \sin\theta}. \tag{2.7}$$

Bandwidths B_c of tens of MHz are easily available and resolutions r_g better than 25 m can be easily obtained. Recently developed SAR systems such as the COSMO-SkyMed SAR constellation operated by the Italian Space Agency (ASI) can reach a spatial resolution of 0.5 m.

2.5.1.2 Along-track spatial resolution

The resolution along track (azimuth direction) for a SLAR is essentially determined by the length L_a of the antenna in the azimuth direction. For an antenna of length L_a, which is wide if compared to its operating wavelength λ, the angular resolution Θ_a in the azimuth direction expressed in radians is:

$$\Theta_a = \frac{\lambda}{L_a}. \tag{2.8}$$

If R denotes the height of the platform with respect to the ground, the along-track resolution r_a is:

$$r_a = \frac{\lambda}{L_a} R. \tag{2.9}$$

For a SLAR orbiting the Earth, considering typical operating parameters, the available azimuth resolution would be of the order of Kilometers and thus absolutely insufficient for most of applications. Fortunately, a solution exists and is represented by SAR.

2.5.1.3 Synthetic aperture radar (SAR)

The basic idea to increase the azimuth resolution is to *synthesize* a longer antenna by exploiting the movement along the track of the platform. Although not immediate to verify, this is possible and the result is that the length of the synthesized antenna is given by the distance traveled by the platform for the time that a given spot on the ground is irradiated by the radar.

In particular, it can be demonstrated [220] that, under acceptable simplifications, by properly processing the SAR signal, the azimuth resolution that can be obtained is:

$$r_a = \frac{L_a}{2}. \tag{2.10}$$

This result is of considerable importance. The across-track resolution is dependent only on the real length of the antenna in the along-track direction, L_a, and in particular it is independent of the slant range, platform altitude, and operating wavelength. In addition, SAR can operate at any altitude without variations in the azimuth resolution from space or from aircrafts.

Another important relation that can be derived concerns the swath width S. By considering the vertical length, L_v, of the antenna corresponding to the across-track direction, analogous to (2.8), we can assume that the angular resolution Θ_v in the across-track direction is expressed in radians as:

$$\Theta_v = \frac{\lambda}{L_v}. \tag{2.11}$$

The swath width S can be thus approximated by:

$$S = \frac{\lambda R}{L_v \cos \theta}. \tag{2.12}$$

Eventually, we can define the ground range and azimuth resolution cells in which the signal coming from the targets of the scene is recorded as shown in Figure 2.19. In fact, by recalling that the resolution cell in the range direction can be expressed by (2.7), the swath width S results divided in a number $N_g = S/r_g$ of elementary resolution cells. The number of cells across the swath N_g and the number N_a of lines acquired by the satellite with resolution r_a in the azimuth direction, determine the size of the radar image in pixels as $N_g \cdot N_a$.

2.5.1.4 SAR images and speckle

The imaging process reported in the previous section is responsible for image formation of SAR *raw data*. Indeed, raw data appear quite uncorrelated because they still contain the effect of the modulation by the chirp signal. For instance, because of this effect, a point target results dispersed in many resolution cells in both range and azimuth direction. To eliminate this unwanted effect in the range direction, a deconvolution is performed on the acquired signal by correlating it with the chirp signal [220]. Because of processing constraints this operation is usually performed in the Fourier domain on the transformed signals and then returning back to the original domain. In fact, in this domain, correlation is obtained by simply multiplying the transformed signals. In addition, the Fourier transform of the chirp signal is always the

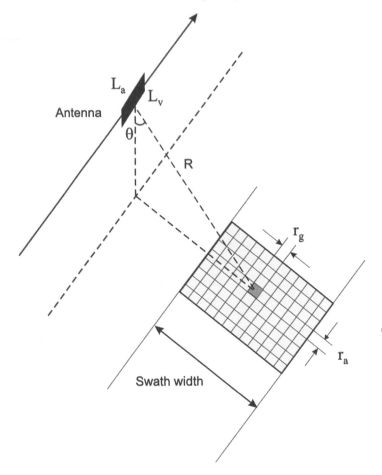

FIGURE 2.19: SAR image formation. Along-track resolution r_a, across-track resolution r_g, and swath width S define the characteristics and the size of the acquired SAR image.

same and this reduces computing time. The result of this operation is to concentrate the originally dispersed energy of the scatterers in the range direction in a few pixels and thus to spatially *compress* it.

A similar deconvolution is then performed in the azimuth direction by compressing data also along track. The result is the production of the so-called *SAR compressed data* that are more familiar to users and in which objects of the scene appear re-focused. This is the reason why the compression process of SAR is often indicated as *focusing*.

A peculiar difference between the appearance of a SAR and an optical image is due to the presence in SAR images of a characteristic noise denoted as speckle. Speckle is caused by the presence of many elemental scatterers with a random distribution within a resolution cell. The coherent sum of their amplitudes and phases results in strong fluctuations of the backscattering from resolution cell to resolution cell. Consequently, the intensity and the phase in the final image are no longer deterministic, but follow instead an exponential and uniform distribution, respectively [198]. Speckle appears as a signal-dependent multiplicative noise whose intensity is thus related to the intensity of the signal, that is, it is stronger on bright areas than on darker ones.

To mitigate speckle effects, *multilooking* is utilized, which is basically an incoherent averaging of the intensity image [77]. Although multilooking reduces the image resolution, it strongly improves the interpretation of SAR images. With modern high-resolution SAR systems, the effect of speckle tends to weaken since the number of elemental scatterers within a resolution cell decreases. Also digital adaptive spatial filters are commonly adopted to reduce the speckle noise [42].

2.5.2 SAR Sensors

A huge amount of literature is available on SAR systems and missions. We report here only a brief summary. Interested readers can refer to the recent papers of Ouchi [201] and Moreira *et al.* [193], where an exhaustive and recent bibliography can be found.

The SeaSat mission of NASA/JPL brought to the launch of the first civilian spaceborne satellite hosting an imaging radar instrument (SAR) in 1978. Notwithstanding the brief life of the satellite (less than four months), the mission fully demonstrated the potentialities of SAR observation technology, generating great interest in satellite active microwave remote sensing. Indeed, SeaSat imaging SAR settled the basis of satellite oceanography and proved the viability of imaging radar for planetary studies by providing extremely interesting images of the Earth surface at a resolution of 25 m with 4-look averaging.

The next spaceborne SAR after SeaSat was ERS-1 in 1991 followed by ERS-2 in 1995. ERS was the first ESA program in Earth observation with the overall objectives to provide environmental monitoring, in particular in the microwave spectrum.

During the 13-year interval between SeaSat and ERS-1, many experiments were developed with airborne SARs and Shuttle Imaging Radar (SIR) series. The SIR-A mission occurred in 1981 with a L-band HH-polarization SAR onboard, similar to SeaSat-SAR. The SIR-B mission followed in 1984 with SAR operating at the same frequency and polarization as those of SIR-A, but varying incidence angles by a mechanically steered antenna. The SIR missions continued in 1994 with the SIR-C/X-SAR by operating at multifrequency X-, C-, and L-bands with a fully polarimetric mode. The Shuttle Radar Topography Mission (SRTM) in 2000 carried X- and C-band main antennas on the cargo bay and a second outboard antenna separated by a

TABLE 2.3: Spaceborne and Shuttle-borne SARs for Earth observation. Resolution is the maximum available spatial resolution in the unit of meters in the azimuth (single-look) and range directions; dual and quad imply the two and four polarization modes, respectively.

Satellite	Agency/Country	Year	Band	Res.	Pol
SeaSat-SAR	NASA/USA	1978	L	6, 25	HH
SIR-A1	NASA/USA	1981	L	7, 25	HH
SIR-B1	NASA/USA	1984	L	6, 13	HH
ERS-1/2	ESA	1991-95	C	5, 25	VV
ALMAZ-1	USSR	1991	S	8, 15	HH
JERS-1 SAR	NASDA/Japan	1992	L	6, 18	HH
SIR-C/X-SAR1	NASA/USA DLR/Germany ASI/Italy	1994	C/L X	7.5, 13 6, 10	quad VV
RadarSat-1	ASC-CSA/Canada	1995	C	8, 8	HH
SRTM1	NASA/USA DLR/Germany	2000	C X	15, 8 8, 19	dual VV
EnviSat-ASAR	ESA	2002	C	10, 30	dual
ALOS-PALSAR	JAXA/Japan	2006	L	5, 10	quad
SAR-Lupe (5)	Germany	2006-08	X	0.5, 0.5	quad
RadarSat-2	ASC-CSA/Canada	2007	C	3, 3	quad
COSMO-SkyMed(6)	ASI/Italy	2007-10	X	1, 1	quad
TerraSAR-X	DLR/Germany	2007	X	1, 1	quad
TanDEM-X	DLR/Germany	2009	X	1, 1	quad
RISAT-1	ISRO/India	2012	C	3, 3	dual
HJ-1-C	China	2012	S	5, 20	VV
Sentinel-1(2)	ESA	2014	C	1.7, 4.3	dual
RadarSat-3(6)	ASC-CSA/Canada	2016	C	1, 3	quad

60 m long mast. With its interferometric system using the two antennas, the SRTM produced a digital elevation model (DEM) at a near-global scale.

An increasing number of spaceborne SARs have been launched recently as can be seen in Table 2.3, and further missions are being planned. Among them we can recall the Sentinel-1 ESA mission and the RadarSat Constellation Mission (RCM) operated by ASC-CSA.

Sentinel-1 is a two satellite constellation, the first of which was launched on 3 April 2014 with the main objectives of land and ocean monitoring. The goal of the mission is to provide C-Band SAR data continuity after the retirement of ERS-2 and the end of the EnviSat mission. The satellites are equipped with a C-SAR sensor, which offers medium- and high-resolution imaging in all weather conditions.

RCM will be composed by a minimum of 3 and up to 6 satellites and is an evolution of the RadarSat program with improved operational use of SAR data and improved system reliability. The overall objective of RCM is to provide C-band SAR data continuity for the RadarSat-2 users, as well as adding a new series of applications that will be made possible through the constellation approach.

As a final consideration, the general trends of SAR systems are that the spatial resolution is becoming finer and different beam modes are becoming available including high-resolution spotlights and wide-swath scan modes with coarser resolution. The conventional single-polarization mode is becoming a dual or fully polarimetric mode.

2.5.3   LiDAR

LiDAR (*Light Detection And Ranging*) is an instrument that collects, analyzes, and records the signal (light) emitted by a laser that is backscattered by objects or by the atmosphere. It has found application in several fields, including RS. To this purpose, it is utilized to measure distances and thus to create digital elevation models or to characterize atmosphere or vegetation. Airborne laser-scanning technology is capable of providing extremely accurate, detailed 3-D measurements of the ground, vegetation, and buildings. The accuracy in measuring elevation can be very high in open, flat areas (10 cm) while may degrade on vegetated areas still maintaining in the range of 50 cm depending on the density of canopies and on the spacing of laser shots. The speed and accuracy of LiDAR scanning makes this technique the most efficient and accurate way to map large areas in a very short time. LiDAR scanning has also found applications in the framework of environmental hazards to map inundated areas, earthquake risk zones, and landslides.

Instruments are usually hosted on a plane or a helicopter. Although several variants are possible, airborne LiDAR technology is based on four main subsystems: a laser emitter-receiver scanning unit fixed to the aircraft platform; global positioning system (GPS) units on the aircraft and on the ground; an inertial measurement unit (IMU) attached to the scanner, which measures attitude parameters of the aircraft; a computer to control the system and store data.

Commercial systems commonly used are discrete-return, small-footprint systems, that is, LiDAR where reflected energy is quantized at amplitude intervals and is recorded at precisely referenced points in time and space. The term *small footprint* indicates that the laser beam diameter at ground level is typically in the range $20 - 100$ cm. The laser scanner shoots up to 250,000 pulses per second to the ground and measures the return time. Starting from the time, the distance each pulse traveled from scanner to ground is computed. The GPS and IMU units determine the location and attitude of the laser scanner as the pulses are emitted, and an exact coordinate is calculated for each point. The laser scanner is a whisk broom sensor. It typically adopts an oscillating mirror or a rotating prism to sweep the light pulses across the swath. Terrain areas are scanned with parallel lines by exploiting the movement of the aerial platforms.

The laser scanner makes 3-D measurements of the ground surface, vegetation, roads, and buildings available. After the flight, software calculates the final data points by using the location information and laser data.

2.6   Concluding Remarks

The basic concepts of remote sensing have been introduced in this chapter with the aim of making the reader familiar with some topics that are relevant to fusion algorithms described in the following chapters. Principles and physical entities involved on remote sensing are first examined by focusing on basic parameters of sensors. In fact, these parameters determine the characteristics of the acquired images that are relevant to pansharpening. In this perspective, importance is given to the concept of spatial resolution by presenting the geometric definition in relation to the definition that can be formulated in terms of point spread function. The PSF is taken as the starting point to introduce and discuss the modulation transfer function that is one of the most relevant points in pansharpening and in the description of any optical acquisition system. Some relevance is also given to the acquisition modalities by describing and comparing whisk broom and push broom sensors. In particular, advantages of the latter with respects to the former are highlighted.

Afterward, optical sensors were presented by distinguishing between reflected radiance and thermal imaging sensors. A classification of sensors based on their spatial resolution is proposed and a review of very high-spatial resolution sensors orbiting on satellites is presented by focusing on those usually involved in pansharpening.

Active sensors were then briefly introduced by describing the principles of synthetic aperture radar and LiDAR. Since such sensors are producing ever-increasing spatial and spectral resolution data, respectively, their importance in image fusion is expected to increase in the near future.

Chapter 3

Quality Assessment of Fusion

3.1   Introduction

Quality assessment of remote sensing image fusion has been, and still is, the object of extensive research activities. Unlike fusion aimed at vision/detection/recognition purposes, like surveillance or for military applications, in which quality is mainly related to a subjective perception that may be embodied by several statistical single-image indices, like contrast, gradients, entropy, and so on, remote sensing image fusion requires the definition of more stringent and quantitative measurements that involve both original images and fusion product images. Since the target of fusion is generally unavailable and thus cannot be used as reference, several protocols of quality evaluation have been developed to overcome the lack of a reference.

Given two datasets, fusion aims at producing a third dataset that inherits some of the properties of its components. The quality of a fusion product should be related to its consistency with the component datasets according to the different properties that are inherited. In most cases, one of the datasets involved in the fusion process features a spectral diversity. The different response of the imaged scene in different intervals of wavelengths is responsible for the spectral information. After fusion, at least in principle, it should be possible to recover the spectral information of the original dataset having spectral diversity. If none of the datasets exhibits a spectral diversity, typical case is the fusion of thermal images with mono-band visible images, the primary information to be preserved after fusion is the radiometry of the thermal image, which can be related to a surface temperature map. The secondary information is the geometry of the visible images, which is accommodated in the low-resolution, or *smooth*, thermal image, after its coregistration with the visible image.

This chapter provides an overview of the materials and methods encountered for quality definition and assessment of remote sensing image fusion, of which pansharpening is the most relevant example. The concept of spectral and spatial quality and the most widely used protocols and statistical indices used for their measurements are presented and discussed. The extension of methods suitable for multispectral pansharpening to the case of hyperspectral pansharpening is discussed. The latter may be challenging, because most of the bands of a hyperspectral image do not overlap to the spectral interval of the enhancing panchromatic image. The borderline case is thermal and visible fusion, in which not only a spectral overlap is missing, but also the imaging mechanisms in the two wavelengths are different, though in both cases photons of electromagnetic radiations are counted. Ultimately, fusion of optical and SAR data is addressed in terms of quality. Here, the geometries of acquisition and the imaged physical characteristics are thoroughly different and the concept of consistency must be properly envisaged.

3.2   Quality Definition for Pansharpening

Pansharpening is a branch of data fusion that is receiving an ever increasing attention from the remote sensing community. New-generation spaceborne imaging sensors operating in a variety of ground scales and spectral bands provide huge volumes of data having complementary spatial and spectral resolutions. Constraints on the signal-to-noise ratio (SNR) impose that the spatial resolution must be lower, if the requested spectral resolution is higher. Conversely, the highest spatial resolution is obtained by a (Pan) panchromatic image, in which spectral diversity is missing. The trade-off of spectral and spatial resolution makes it desirable to perform a spatial resolution enhancement of the lower-resolution multispectral (MS) data or, equivalently, to increase the spectral resolution of the dataset having a higher ground resolution, but a lower spectral resolution; the most notable example is constituted by a unique Pan image bearing no spectral information.

To pursue this goal, an extensive number of methods have been proposed in the literature during the last two decades. Most of them follow a general protocol, that can be summarized in the following two key points: (1) extract

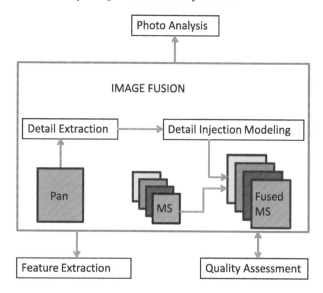

FIGURE 3.1: Pansharpening of multispectral imagery: a synopsis.

high-resolution geometrical information of the scene, not present in the MS image, from the Pan image; (2) incorporate such spatial details into the low-resolution MS bands, interpolated to the spatial scale of the Pan image, by properly modeling the relationships between the MS bands and the Pan image. Figure 3.1 outlines the flow inside the fusion block and highlights possible connections with outer blocks.

Progress in pansharpening methods has been substantially motivated by advances in spaceborne instruments. All instruments launched during the last decade exhibit a ratio of scales between Pan and MS equal to four, instead of two, like in earlier missions, together with the presence of a narrow band in the blue (B) wavelengths and a broadened bandwidth of Pan, also enclosing the near-infrared (NIR) wavelengths. While the change in scale ratios has not substantially influenced the development of fusion methods, the presence of the B band, allowing natural or "true"-color display, and of a Pan image that embraces NIR, but not B, to avoid atmospheric scattering, has created significant problems to earlier methods, thereby dramatically motivating the development both of quality assessment tools for pansharpened data, and of alternative fusion methods yielding better quality than earlier ones. In fact, such methods as intensity–hue–saturation (IHS) [61], Brovey transform (BT) [122], and principal component analysis (PCA) [232] provide superior visual high-resolution multispectral images but ignore the requirement of high-quality synthesis of spectral information [261]. While these methods are useful for visual interpretation, high-quality synthesis of spectral information is very important for most remote sensing applications based on spectral signatures, such as lithology and soil and vegetation analysis [106].

Quality assessment of pansharpened MS images is a much debated topic [66, 261, 278, 33, 158, 37, 32, 148]. The most crucial problem is that if quality is evaluated at the highest resolution of the Pan image, the measured spectral and spatial qualities may follow opposite trends, with the paradox that the least spectrally distorted fused image is that obtained when no spatial enhancement is introduced. The so-called spectral-spatial distortion trade-off occurs because of incorrect definitions and measurements of either spectral or spatial distortion [246]. In the absence of shortcomings, like performing fusion on spatially degraded MS and Pan data in order to evaluate the quality of fusion, the reference for spectral quality is the original MS data to be fused, while the reference of spatial quality is the Pan image. The majority of methods perform a direct comparison between data before and after fusion and this originates the trade-off. To overcome this inconvenience, some authors have introduced new definitions of distortion measurements [32, 148], such that they do not depend on the unavailable true high-resolution MS data, but would measure zero distortions if such data were hypothetically available.

3.2.1 Statistical Quality/Distortion Indices

Quality indices and/or distortion measurements have been defined in order to measure the similarity between images, either scalar or vector [169], as required by the various protocols. This review is limited to indices that are established in the literature as providing results consistent with photoanalysis of pansharpened products. Under this perspective, quality measures based on Shannon's entropy, or auto-information, though used sometimes, have never given evidence of being suitable for this task.

3.2.1.1 Indices for scalar valued images

- Mean bias ($\Delta\mu$): given two scalar images A and B with means $\mu(A)$ and $\mu(B)$, approximated by averages $\bar{A}$ and $\bar{B}$, the mean bias is defined as:

$$\Delta\mu \triangleq \mu(A) - \mu(B) \qquad (3.1)$$

$\Delta\mu$ is a distortion; hence its ideal value is zero.

- Mean square error (MSE): MSE between A and B is defined as:

$$\mathrm{MSE} \triangleq E[(A - B)^2] \qquad (3.2)$$

in which the expected value is approximated by a spatial average. MSE is a distortion, whose ideal value is zero, if and only if $A = B$.

- Root mean square error (RMSE): RMSE between A and B is defined as:

$$\mathrm{RMSE} \triangleq \sqrt{E[(A - B)^2]} = \sqrt{\mathrm{MSE}} \qquad (3.3)$$

in which the expected value is approximated by a spatial average. RMSE is a distortion, whose ideal value is zero, if and only if $A = B$.

- Normalized root mean square error (NRMSE): NRMSE between A and B is defined as:

$$\text{NRMSE} \triangleq \frac{\sqrt{E[(A-B)^2]}}{\mu(B)} \qquad (3.4)$$

in which the expected value is approximated by a spatial average. NRMSE is a distortion; hence its ideal value is zero, if and only if $A = B$. NRMSE is usually expressed in percentage (NRMSE%), especially when A is the test image and B is the reference image.

- (Cross)correlation coefficient (CC): CC between A and B is defined as:

$$\text{CC} \triangleq \frac{\sigma_{A,B}}{\sigma_A \sigma_B} \qquad (3.5)$$

where $\sigma_{A,B}$ is the covariance between A and B, given by $E[(A - \mu(A))(B - \mu(B))]$, and σ_A is the standard deviation of A given by $\sqrt{E[(A-\mu(A))^2]}$. In the same way, $\sqrt{E[(B-\mu(B))^2]}$ represents the standard deviation of B. CC takes values in the range [-1,1]. CC = 1 means that A and B differ only by a global mean offset and gain factor. CC = -1 means that B is the negative of A (A and B still may differ by a gain and an offset). Being CC a similarity index, its ideal value is one.

- Universal image quality index (UIQI) [262] measures the similarity between two scalar images A and B and is defined as:

$$Q = \frac{4\sigma_{A,B} \cdot \bar{A} \cdot \bar{B}}{(\sigma_A^2 + \sigma_B^2)[(\bar{A})^2 + (\bar{B})^2]} \qquad (3.6)$$

in which $\sigma_{A,B}$ denotes the covariance between A and B, $\bar{A}$ and $\bar{B}$ are the means, and σ_A^2 and σ_B^2 the variances of A and B, respectively. Eq. (3.6) may be equivalently rewritten as a product of three factors:

$$Q = \frac{\sigma_{A,B}}{\sigma_A \cdot \sigma_B} \cdot \frac{2\bar{A} \cdot \bar{B}}{[(\bar{A})^2 + (\bar{B})^2]} \cdot \frac{2\sigma_A \cdot \sigma_B}{(\sigma_A^2 + \sigma_B^2)}. \qquad (3.7)$$

The first one is the correlation coefficient (CC) between A and B. The second one is always ≤ 1, from Cauchy-Schwartz inequality, and is sensitive to bias in the mean of B with respect to A. The third term is also ≤ 1 and accounts for changes in contrast between A and B. Apart from CC which ranges in $[-1, 1]$, being equal to 1 *iff* $A = B$, and equal to -1 *iff* $B = 2\bar{A} - A$, i.e., B is the negative of A, all the other terms range in $[0, 1]$, if $\bar{A}$ and $\bar{B}$ are nonnegative. Hence, the dynamic range of Q is [-1,1] as well, and the ideal value $Q = 1$ is achieved *iff* $A = B$ for all pixels. To increase the discrimination capability of the three factors in (3.7), all statistics are calculated on suitable $N \times N$ image blocks and the resulting values of Q averaged over the whole image to yield a unique global score.

3.2.1.2   Indices for vector valued images

- Spectral angle mapper (SAM). Given two spectral vectors, $\mathbf{v}$ and $\hat{\mathbf{v}}$, both having L components, in which $\mathbf{v} = \{v_1, v_2, \cdots, v_L\}$ is the original spectral pixel vector $v_l = A^{(l)}(m, n)$ while $\hat{\mathbf{v}} = \{\hat{v}_1, \hat{v}_2, \cdots, \hat{v}_L\}$ is the distorted vector obtained by applying fusion to the coarser resolution MS data, i.e., $\hat{v}_l = \hat{A}^{(l)}(m, n)$, the spectral angle mapper (SAM) [273] denotes the absolute value of the spectral angle between the two vectors:

$$\text{SAM}(\mathbf{v}, \hat{\mathbf{v}}) \triangleq \arccos\left(\frac{<\mathbf{v}, \hat{\mathbf{v}}>}{||\mathbf{v}||_2 \cdot ||\hat{\mathbf{v}}||_2}\right). \qquad (3.8)$$

  $\text{SAM}(A, B)$ is defined according to (3.8) as $\text{E}[\text{SAM}(a, b)]$ where a and b denote the generic pixel vector element of MS images A and B, respectively. SAM is usually expressed in degrees and is equal to zero when images A and B are spectrally identical, i.e., all pixel vectors differ only by their moduli between A and B.

- *Relative average spectral error* (RASE). RASE [217] is an error index that offers a global indication of the quality of a fused product. It is given by:

$$\text{RASE} \triangleq \frac{100}{\sum_{l=1}^{L} \mu(l)} \sqrt{L \sum_{l=1}^{L} \text{MSE}(l)} \qquad (3.9)$$

  where $\mu(l)$ is the mean (average) of the lth band, and L is the number of bands. Low values of RASE indicate similarity between multispectral data.

- *Relative dimensionless global error in synthesis* (ERGAS). ERGAS [260], another global error index and an improvement of RASE (3.9), is given by:

$$\text{ERGAS} \triangleq 100\frac{d_h}{d_l} \sqrt{\frac{1}{L} \sum_{l=1}^{L} \left(\frac{\text{RMSE}(l)}{\mu(l)}\right)^2} \qquad (3.10)$$

  where d_h/d_l is the ratio between pixel sizes of Pan and MS, e.g., 1/4 for IKONOS and QuickBird data, $\mu(l)$ is the mean (average) of the lth band, and L is the number of bands. Low values of ERGAS indicates similarity between multispectral data.

- Q4 is a multispectral extension of UIQI, suitable for images having four spectral bands, introduced by three of the authors for quality assessment of Pansharpened MS imagery [33]. For MS images with four spectral bands, let a, b, c, and d denote the radiance values of a given image pixel in the four bands, typically acquired in the B, G, R, and NIR wavelengths. Q4 is made up of different factors accounting for correlation, mean bias, and contrast variation of each spectral band, as well as of spectral angle. Since the modulus of the hypercomplex correlation coefficient (CC) measures the alignment of spectral vectors, its low value may detect when radiometric distortion is accompanied by spectral distortion. Thus, both radiometric and spectral distortions may be encapsulated in a unique parameter. Let

$$\begin{aligned} \mathbf{z}_A &= a_A + \mathbf{i}b_A + \mathbf{j}c_A + \mathbf{k}d_A \\ \mathbf{z}_B &= a_B + \mathbf{i}b_B + \mathbf{j}c_B + \mathbf{k}d_B \end{aligned} \tag{3.11}$$

denote the 4-band reference MS image and the fusion product, respectively, both expressed as quaternions or hypercomplex numbers. The Q4 index is defined as:

$$Q4 \triangleq \frac{4|\sigma_{z_A z_B}| \cdot |\bar{\mathbf{z}}_A| \cdot |\bar{\mathbf{z}}_B|}{(\sigma_{z_A}^2 + \sigma_{z_B}^2)(|\bar{\mathbf{z}}_A|^2 + |\bar{\mathbf{z}}_B|^2)}. \tag{3.12}$$

Eq. (3.12) may be written as product of three terms:

$$Q4 = \frac{|\sigma_{z_A z_B}|}{\sigma_{z_A} \cdot \sigma_{z_B}} \cdot \frac{2\sigma_{z_A} \cdot \sigma_{z_B}}{\sigma_{z_A}^2 + \sigma_{z_B}^2} \cdot \frac{2|\bar{\mathbf{z}}_A| \cdot |\bar{\mathbf{z}}_B|}{|\bar{\mathbf{z}}_A|^2 + |\bar{\mathbf{z}}_B|^2} \tag{3.13}$$

the first of which is the modulus of the hypercomplex CC between $\mathbf{z}_A$ and $\mathbf{z}_B$ and is sensitive both to loss of correlation and to spectral distortion between the two MS datasets. The second and third terms, respectively, measure contrast changes and mean bias on all bands simultaneously. Ensemble expectations are calculated as averages on $N \times N$ blocks. Hence, Q4 will depend on N as well. Eventually, Q4 is averaged over the whole image to yield the *global* score index. Alternatively, the minimum attained by Q4 over the whole image may represent a measure of *local* quality. Q4 takes values in [0,1] and is equal to 1 only when $A = B$. Q4 has been recently extended to deal with images, whose number of bands is any power of two [116].

.

3.2.2   Protocols Established for Pansharpening

The use of one or another statistical index is by itself not sufficient to guarantee that the quality evaluation is correct and thus not misleading. What is missing is the definition of a suitable operational protocol, such that the spectral and spatial consistencies can be reliably measured. Before the seminal paper by Wald *et al.* [261], a definition of quality was missing and empirical indices were coupled with even more empirical criteria in order to measure a quality that in most cases was only spatial. As an example, Chavez Jr. *et al.* [66] report the percentage of changed pixel (within a $(-\epsilon, \epsilon)$ tolerance) between sharpened and interpolated bands as a figure of merit of pansharpening. Following the authors' guidelines [66], such a percentage should be as small as possible, because too high a value would indicate an overenhancement, with consequent introduction of spatial artifacts. However, anyone could notice that a zero percentage would indicate the absence of spatial enhancement, at least of an enhancement outside the tolerance interval. Thus, the threshold ϵ would be crucial for the significance of the quality criterion based on changed/unchanged pixels. A further remark is that the optimal percentage of changed pixels depends on the landscape and cannot be *a priori* set equal to a few percent for any scene.

3.2.2.1   Wald's protocol

A general protocol usually accepted in the research community for quality assessment of fused images was first proposed by Wald *et al.* [261] and rediscussed in Ranchin *et al.* [216] and in Thomas *et al.* [246]. Such a protocol relies on three properties that the fused data have to satisfy.

The first property, known as *consistency*, requires that *any fused image $\hat{A}$, once degraded to its original resolution, should be as identical as possible to the original image A.* To achieve this, the fused image $\hat{A}$ is spatially degraded to the same scale of A, thus obtaining an image $\hat{A}^*$. $\hat{A}^*$ has to be very close to A. It is worthwhile that consistency measures spectral quality after spatial enhancement and is a condition necessary but not sufficient to state that a fused image possesses the necessary quality requirements, that is, both spectral and spatial quality.

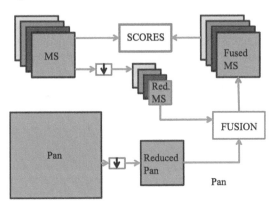

FIGURE 3.2: Procedure to assess Wald's synthesis property.

The second property, known as *synthesis* states that *any image $\hat{A}$ fused by means of a high-resolution (HR) image should be as identical as possible to the ideal image A_I that the corresponding sensor, if existent, would observe at the resolution of the HR image.* Images are regarded here as scalar images, that is, one spectral band of an MS image. Similarity is measured by statistics of scalar pixels between a fused image and its ideal HR reference. Besides scalar similarity indices between individual bands of the MS image, the synthesis property is checked on the plurality of spectral bands constituting an MS image, in order to check the multispectral properties of the MS image, that is, of the whole set of fused bands: *the multispectral vector of images $\vec{A}$ fused by means of a high-resolution (HR) image should be as identical as possible to the multispectral vector of ideal images $\vec{A}_I$ that the corresponding sensor, if existent, would observe at the spatial resolution of the HR image.* This second part of synthesis property is also known as third property. Both the synthesis properties may not generally be directly verified, since $\vec{A}_I$ is generally not available. Therefore, synthesis is usually checked at degraded spatial scales according to the scheme of Figure 3.2. Spatial degradation is achieved by means of proper lowpass filtering followed by decimation by a factor equal to the scale ratio of Pan to MS datasets. The multispectral image $\vec{A}^*$ and the panchromatic image P* are created from the original sets of images $\vec{A}$ and Pan. Pan is degraded to the resolution of the multispectral image and $\vec{A}$ to a lower resolution depending on the scale ratio for which the fusion is assessed. The fusion method is applied to these two sets of images, resulting into a set of fused images at the resolution of the original MS image. The MS image serves now as reference and the second and third properties can be tested. It is noteworthy that fulfillment of the synthesis properties is a condition both necessary and sufficient, provided that the similarity check performed at the degraded spatial scale is consistent with the same check if it were hypothetically performed at the full scale; in other words, if the quality observed for the fused products is assumed to be close to the quality that would be observed for the fused products at the full scale. This point is crucial, especially for methods employing digital filters to analyze the Pan image. In fact, whenever simulations are carried out at degraded spatial scale, the lowpass filter of the fusion methods is cascaded with the lowpass filter used for decimation. Hence, fusion at full scale uses the former only; fusion at degraded scale uses the cascade of the former with the latter, that is, uses a different filter [14]. This explains why methods providing acceptable spatial enhancement at degraded scale yield poor enhancement when they are used at full scale [158].

3.2.2.2 Zhou's protocol

As an alternative to Wald's protocol, the problem of measuring the quality of fusion may be approached at the full spatial scale without any spatial degradation [278]. The spectral and spatial distortions are separately evaluated from the available data, that is, from the original low-resolution MS bands and high-resolution Pan image. The spectral distortion is calculated for each band as the average absolute difference between the fused band and the interpolated original bands, while the spatial quality is measured by the correlation coefficient (CC) between the spatial details of each of the fused MS bands and those of the Pan image; such details are extracted by means of a Laplacian filter and the outcome spatial CC (sCC) should in principle be as close to one as possible, though no evidence of that is given in the original paper (Zhou, Civco, and Silander [278]) or in subsequent ones.

3.2.2.3   QNR protocol

The Quality w/ No Reference (QNR) protocol (Alparone *et al.* [32]) calculates the quality of the pansharpened images without requiring a high-resolution reference MS image. QNR comprises two indices, one pertaining to spectral and the other to spatial distortion. The two distortions may be combined together to yield a unique quality index. However, in many cases they are kept separate. Both spectral and spatial distortion are calculated through similarity measurements of couples of scalar images performed by means of UIQI.

The spectral distortion D_λ is calculated between the low-resolution MS images and the fused MS images. Hence, for determining the spectral distortion, two sets of inter-band UIQI values are calculated separately at low and high resolutions. The differences of corresponding UIQI values at the two scales yields the spectral distortion introduced by the pansharpening process. Thus, spectral distortion can be represented mathematically as:

$$D_\lambda = \sqrt[p]{\frac{1}{N(N-1)} \sum_{l=1}^{N} \sum_{r=1, r \neq l}^{N} \left| Q(\tilde{M}_l, \tilde{M}_r) - Q(\hat{M}_l, \hat{M}_r) \right|^p} \qquad (3.14)$$

where, $\tilde{M}_l$ represents the low-resolution lth MS band, $\hat{M}$ the pansharpened MS band, $Q(A, B)$ represents UIQI between A and B and N is equal to the number of MS bands. The exponent p is an integer possibly chosen to emphasize large difference values. By default p is set equal to one.

The spatial distortion D_S is determined by calculating the UIQI between each MS band and the Pan image degraded to the resolution of MS and again between fused MS and full-resolution Pan. The difference between the two values yields the spatial distortion:

$$D_s = \sqrt[q]{\frac{1}{N} \sum_{l=1}^{N} \left| Q(\tilde{M}_l, P_L) - Q(\hat{M}_l, P) \right|^q} \qquad (3.15)$$

in which P_L denotes the Pan image degraded to the resolution of MS and P the high-resolution Pan image. The exponent q is one by default.

The rationale of QNR protocol is the following:

1. The inter-relationships (measured by UIQI) between couples of the low-resolution MS bands should not change with resolution, i.e., once the MS image has been pansharpened.

2. The relationships between each MS band and a low-resolution version of the Pan image should be identical to those between each pansharpened MS band and the full-resolution Pan image.

3. Differences in similarity values computed at low and high resolution measure the distortion, either spectral (MS-MS) or spatial (MS-Pan).

An earlier version of QNR employed mutual information instead of UIQI in order to measure the similarity between two grayscale images [30].

3.2.2.4 Khan's protocol

The Khan protocol [148] borrows the consistency property from Wald's protocol, the matching of highpass spatial details from Zhou's protocol and the definition of spectral distortion from the QNR protocol, in order to define separate spectral and spatial quality indices at full scale. The unifying framework is that the pansharpened image can be regarded as the sum of a lowpass term, corresponding to the original interpolated low-resolution MS image, and a highpass term, corresponding to spatial details extracted from Pan that have been injected. Such components are extracted by filtering the fused image with a bank of digital filters, the lowpass filter matching the shape of the MTF of the corresponding spectral channel, the highpass filter matching the one's complement of the MTF, that is, the highpass filter is an allpass filter minus the lowpass filter. Spectral quality is evaluated on such lowpass component, while spatial quality on the highpass one.

In a practical implementation, Gaussian models of the MTFs of instrument MS channels provide the lowpass components. Such components are decimated and the similarity with the original low-resolution MS data is measured by means of Q4 [33], or any other similarity index for vector data if the number of bands is not four. The highpass components are simply given by the fused image minus the lowpass-filtered image before decimation. Highpass details of Pan are extracted by means of the same highpass filter as MS. UIQI [262] is calculated between details of each fused MS band and details of Pan. The UIQIs of each band are averaged together to yield $UIQI_H$, which measures the similarity with Pan of spatial structures at high resolution. The procedure of highpass details extraction and matching is repeated for the original low-resolution MS image and a spatially degraded version of Pan, achieved through a selective filter with 1:4 frequency cutoff. Thus, an index measuring the similarity with Pan of spatial structures at low resolution, $UIQI_L$, is obtained. Eventually, the (absolute) difference of $UIQI_L$ and $UIQI_H$ is taken as a measurement of spatial distortion.

3.2.3 Extension to Hyperspectral Pansharpening

For the case of hyperspectral pansharpening, which will be addressed in Chapter 8, the definitions of spectral and spatial consistency measurements are substantially identical to those of multispectral pansharpening. However, since most of spectral bands are not overlapped with the wavelength interval of Pan, in tests carried out at full scale, that is, the spatial scale of Pan, the spectral consistency of fusion product with the low-resolution hyperspectral data is more significant than the spatial consistency, for example, as measured by QNR or Khan's protocol. The latter is particularly crucial, given the high number of variables that are involved (MTFs of all spectral channels). Besides spectral consistency measurements at full scale, Wald's synthesis property may be checked at the degraded spatial scale. Again, the spatial degradation would in principle require the MTFs of all spectral channels [14]. In this case, a plot of NMRSE (3.4) versus either band index or wavelength is capable of highlighting the trend in fusion accuracy with the wavelength, for example, as a function of the bandwidth of the Pan instrument [17].

3.2.4   Extension to Thermal V-NIR Sharpening

The sharpening of a scene imaged in the thermal infrared wavelength interval by means of a higher-resolution image taken in the visible and/or near-infrared spectral intervals, not necessarily a broadband panchromatic image, is a problem that may be encountered outside remote sensing. In this case, the consistency of the fusion product with the original thermal image is a *radiometric* consistency and may also be measured through a scalar index, for example, UIQI (3.6), for example, by following Wald's consistency property. The spatial quality may be measured, for example, by Zhou's sCC, as described in Section 3.2.2.2.

Whenever a spectral diversity of the thermal band exists, which means that the imaging instrument splits the thermal wavelength interval into a few subintervals and produces an image for each of them, V-NIR-sharpening becomes analogous to multispectral pansharpening and the related indices and protocols apply. However, the total lack of overlap between V-NIR and thermal infrared (TIR or LWIR), together with the weak spectral diversity of thermal bands (color composite display looks grayish and colorless) makes a spectral consistency check on the ensemble of bands to be less stringent than radiometric consistency checks on each band separately. Analogous to the one-band case, the spatial consistency with the enhancing V-NIR image may be measured following any of the full-scale protocols reviewed in this chapter.

3.3   Assessment of Optical and SAR Fusion

The specific topic of optical and SAR fusion deserves an entire section because the problem cannot be assimilated within the quality definition issues of pansharpening. In other words, there is not a reference for optical and SAR image fusion, as for pansharpening. In fact, the reference of a pansharpened image at 1 m, derived from an original MS image at 4 m, is the image one would obtain if the satellite instrument could be placed on an orbit four times lower.

A further problem is that the acquisition geometries of optical scanners and SAR systems makes coregistration of the two datasets extremely challenging, in the presence of some terrain relief. Thanks to the availability of a high-resolution digital elevation model (DEM), a fair coregistration can be achieved, but misregistration effects can impact on the quality of fusion products, such that lacks of consistency may be charged both to the fusion method and to registration inaccuracies and the two contributions cannot be separately measured, for example, by simulating an optical image overlapped to the test SAR image.

Let us split the task of optical and SAR image fusion into two distinct tasks:

1. An optical image, e.g., MS or MS + Pan, is enhanced by means of a coregistered SAR image, which is assumed to be mono-band, single polarization, for sake of simplicity, but in principle it might also feature spectral and or polarization diversity. This case requires the existence and integrity of the optical image, which may be not always available or at least not complete, because of sparse cloud covers. The fusion product is an optical image that inherits some of the properties of the SAR image. The majority of methods in the literature, including Alparone *et al.* [34], which will be reviewed in Chapter 10, pursues this task.

2. A SAR image (available in all weather conditions) is spectrally or spatially or radiometrically enhanced by means of an optical image (MS only, MS + Pan, Pan only). The result is a SAR image that inherits some of the properties of the optical image, e.g., the spectral (color) information, the geometric and texture information. Radiometrically enhanced means that the SNR of the SAR image is increased by the fusion process. Thus, fusion acts as a kind of despeckling filtering controlled by the optical image. The advantage of this dual approach is that a product is always available regardless of the integrity of the optical image. This task has never been pursued in the literature, to the best of the authors' knowledge, because it requires a total neglect of classical pansharpening-like fusion approaches, in favor of an aware and mature capability of catching both applications that may benefit from such fusion products and methodological solutions to combine the heterogeneous datasets.

In the first case, the result is an optical image enhanced by SAR. Thus, the spectral consistency issue foresees that from the fused image something very close to original MS can be retrieved. This is feasible only if the details that have been incorporated to the MS image from the SAR image are *thinner* than the spatial details of the original low-resolution MS image. In this case, the fused MS image can be lowpass filtered and decimated before being matched to the original MS. Any mismatch measured by a proper index is responsible for the loss of spectral quality originated by fusion. Another suitable measure of spectral fidelity to the original is the average spectral angle error between the fusion product and the interpolated MS [34]. A measure of consistency with SAR is less immediately achievable. It would require that the property inherited from SAR, for example, texture, can be recovered from the fused image. In any case, since a target of fusion does not exist, depending on the applications the consistency with SAR may not be required.

In the second case, which has not been well investigated in the literature, the product of fusion is a SAR image, possibly radiometrically enhanced, that is, despeckled, and spectrally enhanced by means of the optical image. In this case, radiometric consistency means that the fused SAR image is multiband and its pixels are vector valued. A norm operation carried out on such vector pixels, typically L_1 or L_2, should produce a scalar image most similar to the original SAR. Conversely, the spectral angle of the *colored* SAR image should match the angle of the enhancing MS image. Thus, the radiometric consistency with SAR and the spectral consistency with MS are assets of a SAR image spectrally enhanced by a coregistered MS observation of the same scene. The possible radiometric enhancement obtained, for example, from the availability of a VHR panchromatic image, which is used to drive a despeckling procedure, requires the definitions of indices, different from those reviewed in this chapter, that are able to measure the quality of a despeckled SAR image [42].

3.4   Concluding Remarks

Quality assessment of fusion products in remote sensing is a crucial task and requires total abandon of quality indices borrowed from computer vision that reflect the contrast, clarity and sharpness of an image to a human observer. In this sense, each index features two arguments, one related to the fused image, the other to the original image. Single-image indices, like average gradients, auto-information or entropy, and so on, are definitely unable to qualify remote sensing fusion products, especially for automated tasks, in which a human observer is not involved. Both single- and two-argument indices are reviewed by Li *et al.* [169], though many of such indices have never been given evidence in scientific publications that they are suitable for remote sensing images.

Whenever pansharpened data are used for automated tasks, like spatial/spectral feature extraction, and generally all applications based on analysis of spectral signatures and modeling of geophysical and biophysical processes, the quality of pansharpened products, trivially depending on the input data and type of fusion algorithm, becomes a crucial issue. While radiometric quality (SNR) of fusion products is generally lower than that of the original MS data, the spatial, or geometric, quality is greater than that of the original MS data and comparable to that of the Pan image. The main novelty of fusion methods developed over the last 12 years is that also spectral quality, or better spectral fidelity to the original low-resolution MS data, may be thoroughly preserved. In other words, the information stemming from the spectral diversity of the original MS image is synthesized at the spatial scale of the fusion product, that is, of the Pan image. Spectral quality, which can be associated to chromatic fidelity to originals in color compositions of three bands at a time, can be measured by means of statistical indices and a suitable protocol, analogously to the spatial/geometric quality. The shortcoming of assessing quality at degraded spatial scale, that is, the synthesis property check of Wald's protocol, does not allow spectral and spatial qualities to be separately measured.

Eventually, we wish to stress once more that the concept of *trade-off* between the spectral and spatial qualities attainable by any fusion methods, introduced by several authors during the last few years, is definitely erroneous and occurs only if the two distortions are not correctly defined and/or measured. Therefore, whenever the design of a new fusion method is driven by some quality index, extreme care should be taken in the choice of the index, and related protocol, to avoid puzzling results. Sadly, the improper definition and use of quality indices and the incorrect application of evaluation protocols, together with the lack of standardized datasets, is responsible for the abnormal growth of remote sensing image fusion methods in the scientific literature over the last decade.

Chapter 4

Image Registration and Interpolation

4.1 Introduction

Image interpolation and registration are two important tools in the field of remote sensing applications, and in addition they are closely related. Actually, the remapping of an acquired image often requires a resampling of the image itself before subpixel translation, rotation, deformation, or warping can be achieved [185]. At the same time, image interpolation is usually the first step of a fusion procedure between two types of remote sensing data having different spatial resolutions, that is, pansharpening. In this case, a preliminary correct registration is mandatory because if the two images are misregistered, the interpolation and the successive fusion procedure cannot produce the desired results [45].

In fact, most of the pansharpening techniques proposed in the literature during the last 25 years require the resampling of the multispectral (MS) image, in order to match its size with the one of the panchromatic (Pan) image, before the introduction of geometric details of the Pan is performed. A general protocol can be defined which is able to accommodate the majority of these methods, so that three main steps have to be employed:

- expand the MS image by proper interpolation, such that the magnified image is pixel-by-pixel overlapped to the Pan image;

- extract the high-resolution spatial details of the scene, not available in the low-resolution MS images, from the Pan image;

- incorporate such a geometrical information into the expanded MS bands, by considering a suitable model of the relationships between the MS and Pan images.

As stated before, a required preprocessing for this set of operations is a registration of MS and Pan images on the same cartographic map, in order to compensate a variety of distortions which arise from the acquisition process, especially if the Pan and MS images have been acquired by different sensors or by different optics. However, the registration process, even if accurate, cannot exactly compensate all the distortions, so that care should be taken in the successive steps, especially in the interpolation procedure.

So far, another issue which has not been adequately evidenced in the literature is the presence of an intrinsic shift between MS and Pan images, if the interpolation is made by means of an odd filter, in images acquired by state-of-art sensors (QuickBird, IKONOS, and so on). This problem will be extensively discussed in this chapter because an inaccurate or even theoretically wrong interpolation strategy can heavily influence the fusion results, so that the performances of a pansharpening technique can be degraded of a factor of 15% or even 20% with respect to the performances of a correct interpolation modality.

The remainder of this chapter is organized as follows. Section 4.2 gives an overview of geometric image correction and registration. Section 4.3 introduces the problem of image interpolation for pansharpening applications, and in particular the importance of piecewise local polynomial kernels is highlighted. Section 4.3.2 recalls the theoretical fundamentals of digital interpolation, starting from continuous-time kernels from which digital kernels can be derived. Section 4.3.5 formalizes the differences between odd and even digital interpolation kernel families and derives the coefficients of digital piecewise polynomial kernels, in particular the cubic one recently proposed in the literature. Section 4.3.6 applies odd and even digital piecewise polynomial kernels to the problem of pansharpening, by showing that only even kernels are suitable for the images acquired by modern sensors. Section 4.3.7 reports practical tests on real images to quantitatively highlight the differences between correct and incorrect interpolation modalities in image pansharpening. Finally, the conclusions are summarized in Section 4.4.

4.2   Image Registration

Multisensor image registration is one of the most important basic processing in most remote sensing applications and especially in pansharpening. Actually, the merging of images acquired from different instruments requires an accurate removing of all the sources of unwanted variations, or distortions, before the data can be processed together, especially if optical and SAR data are considered for the fusion. As an example, we recall the geometric changes caused by the different acquisition parameters and by the deviations of these parameters from their nominal values.

4.2.1   Definitions

Usually, the term *registration* indicates the fitting of the coordinate system of an image to the coordinate system of a reference image of the same area. A related technique which usually precedes the registration process is the *geometric correction*, where the remotely sensed image is transformed so that it acquires the scale and the projection properties of a cartographic map and can be used in a geographic information system (GIS) [181]. The importance of a geometric correction process is becoming more and more pressing as the spatial resolution of the images becomes greater [249]. Some terms usually included in the geometric correction, or *geocorrection*, are *georeferencing*, which means that geographical coordinates are provided for the four corners of the image (but not for the other pixels), *geocoding*, which implies that the image is equipped with all the properties of a map, and *orthorectification*, where the correction process also considers terrain elevation.

In general, the problem of registering two images to be compared or merged arises when the images have been acquired at different times, at different resolutions, by different sensors, or from different viewpoints. Several types of registration methods have been addressed in the literature and briefly reviewed in Sections 4.2.3 and 4.2.5. The most important family is named *point mapping*. For the methods belonging to this family, the registration process is based on a suitable transformation so that the points in the first image can be related with the correspondent points in the other. The choice of the optimal transformation is driven by the types of the variations between the two images [51].

By following the literature, point mapping methods are characterized by four main steps:

- feature detection, which extracts salient objects in the images useful for the successive phase of matching; from these objects the so-called *control points* (CPs) can be derived;

- feature matching, which establishes a correspondence between the features detected in the sensed image and in the reference image, by considering suitable similarity measures;

- transform estimation, which selects the type and the parameters of the mapping function between the two images by using the established feature correspondence and the selected CPs; the estimation of the parameters can be made by considering global information on the entire image or only local information;

- image transformation, which determines the transformed image by means of the mapping function, where resampling techniques based on interpolation are used to determine values in noninteger coordinates.

Evidently, these four steps, and especially the selection of the features of interest, can only be accurately made if the images have been previously corrected geometrically [280].

4.2.2   Geometric Corrections

The integration of images acquired by different sources requires the definition of a common coordinate system, typically the latitude/longitude reference system of a cartographic map, by considering different map projections aimed to preserve shape areas or object bearings [99, 128]. A generic technique that converts the image data into these Earth centric coordinates is called geometric correction, by utilizing for example the global positioning system (GPS). The most accurate form of geometric correction is the orthorectification, which includes the terrain elevation in the correction procedure. The availability of a digital elevation model (DEM) at more resolved scales has made easier the production of orthorectified images also for high-resolution data which are conformed to the map accuracy standards [181, 163, 2], by also utilizing synthetic aperture radar (SAR) data [265].

4.2.2.1   Sources of geometric distortions

The deviations of the geometry of the remotely sensed image with respect to the terrain cannot be avoided in the acquisition process and can considerably differ depending on the platform (aircraft or satellite), on the type of

TABLE 4.1: Geometric error sources internal to the acquisition systems or caused by external factors.

Category	Subcategory	Error Sources
Acquisition system (observer)	Platform (spaceborne or airborne)	Variation of movement Variation in platform attitude (low to high frequencies)
	Sensor (VIR, SAR or HR)	Variation in sensor mechanism (scan rate, scanning velocity) Viewing/look angles Panoramic effect (FOV)
	Measuring instruments	Time variations or drift Clock synchronicity
External factors (observed)	Atmosphere	Refraction and turbulence
	Earth	Curvature, rotation, topographic effects
	Map	Geoid to ellipsoid Ellipsoid to map

sensor (optical or SAR), on the spatial resolution (low or high), and on the total field of view (FOV). However, in general, the sources of distortions can be characterized as to be caused by: (a) the acquisition system, or *observer* (platform, imaging sensor, other measuring instruments), and (b) external factors or *observed* (atmosphere, Earth, deformations of the map projection) [249]. Table 4.1 reports the main sources of distortion in the acquisition process. Due to the different types of acquisition systems, the deviations can be categorized into low, medium or high frequency distortions in relation to the image acquisition time.

The most remarkable sources of geometric errors reported in Table 4.1 can be summarized as [249]:

- instrument errors, mainly due to the distortions in the optical system, nonlinearity of the scanning mechanism, and nonuniform sampling rates. They are generally predictable or systematic, and can be corrected at the ground stations.

- platform instabilities that generate distortions in altitude and attitude. Altitude variations combined with sensor focal length, the Earth's flatness and terrestrial relief can change the pixel spacing. Attitude varia-

tions (roll, pitch, and yaw) can change the orientation and/or the shape
of optical images, but they do not affect SAR acquisitions. If there exist
platform velocity variations, they can change the line spacing or create
line gap and overlaps.

- Earth's effects. The consequence of the Earth's rotation (whose velocity
  is a function of the latitude) is to skew a line scanned image, i.e., it
  generates latitude dependent displacements between image lines. Earth's
  curvature can create variations in pixel spacing. Figure 4.1 shows the
  effects of the Earth's motion on the geometry of a line scanned image.

- other sensor related distortions, as (i) calibration parameter uncertainty
  (focal length and instantaneous field of view (IFOV) for optical images
  and range gated delay for SAR acquisitions) and (ii) panoramic distortions, which are function of the angular FOV of the sensor and affect
  especially sensors with a wide FOV (AVHRR, VIRS) with respect to
  instruments with a narrow FOV (Landsat ETM+, SPOT-HRV). They
  can change the ground pixel sampling along the column.

- atmospheric effects, which are specific to each acquisition time and location and thus often not corrected. They are also negligible for low-to-
  medium resolution images.

- deformations due to the map projection, i.e., the approximation of the
  geoid by a reference ellipsoid, and the projection of the reference ellipsoid
  on a tangent plane.

The geometric corrections of these distortions generally include the following steps: (i) the determination of a relationship between the two coordinate
systems of a map and each image and the two images for the registration; (ii)
the establishment of a set of points in the corrected image that define an image with the desired cartographic properties; and (iii) the estimation of pixel
values to be associated with those points [181]. Such a process is outlined in
Figure 4.2. Substantially, in this process, the determination of suitable models
and mathematical functions is required.

In the case that a physical model is available, then the geometric correction can be performed step-by-step by using a mathematical function for each
distortion, or simultaneously by means of a combined function [249]. The characteristics of the satellite's orbit, the Earth's rotation, and the along-scan and
across-scan sampling rate of the sensor can be used for this goal. If the orbital
geometry of the satellite platforms are sufficiently known, algorithms that uti-
lize the orbital geometry give high accuracy. An example of the utilization of
the orbital geometry model in the geometric correction is depicted in Figure
4.3. Otherwise, orbital models are useful only if the desired accuracy is not
very high, or where maps of the covered areas are not available. Alternatively,
2-D/3-D empirical models (polynomial or rational functions) are considered
for the mapping between terrain and image coordinates.

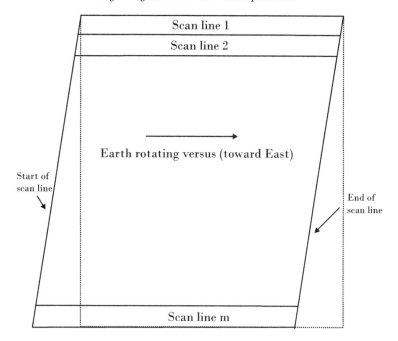

FIGURE 4.1: Effects of Earth rotation on the geometry of a line-scanned image. Due to the Earth's eastward rotation, the start of each swath is moved slightly westward.

4.2.3   Point Mapping Methods: Image Transformations

The orbital geometry model takes into account only geometric distortions caused by factors defined *a priori*. Distortions not available in advance as variations in altitude or attitude of the platform cannot be corrected. An alternative method to the construction of a physical model is to define an empirical model so that the differences between some common point positions on the image and on a reference map can be compared. From these deviations, the image distortions can be estimated and corrected by means of a suitable transformation. If a sufficiently accurate map is not available, the identification of a set of ground control points (GCPs) can be made by using position fixing methods.

4.2.3.1   2-D polynomials

Accurately located control points, which are well defined and easily recognizable features, allow for a relation between the coordinates of the map that is taken as reference and of the sensed image. Denote with (x_n, y_n) the map coordinates of the nth control point, and with (X_n, Y_n) the image coordinates of the correspondent point. The image coordinates denote the pixel position along the scan line (column) and along the column (row), whereas the map

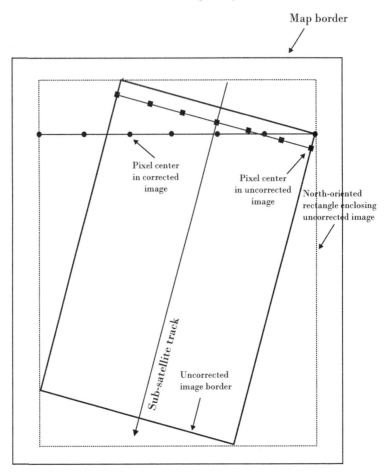

FIGURE 4.2: Example of conversion of map coordinates into image coordinates. Once the image is outlined on the map, by using the map coordinates of the image corners, the map coordinates of the pixel center can be found. The area of the corrected image is shown by the rectangle that encloses the oblique uncorrected image. A least squares transformation allows to obtain the positions of the corners of the enclosing rectangle in map easting and northing coordinates by starting from the coordinates of the corners of the uncorrected image. Once the corners of the corrected image area are obtained in terms of map coordinates, which are expressed in kilometers, the locations of the pixel centers in the corrected image can be determined. Finally, the pixel center positions are converted to image coordinates so that each pixel value can be associated with the pixel position in the corrected image. A null value is assigned to the pixels in the corrected image that do not lie within the area of the uncorrected image.

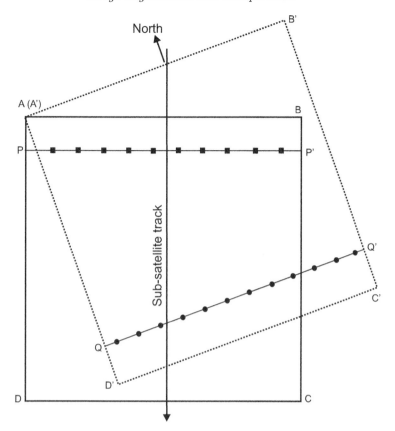

FIGURE 4.3: Utilization of the orbital geometry model in the process of geometric correction of an image. The uncorrected (raw) image is outlined by the solid line (ABCD), whereas the corrected image is outlined by the dashed line (A' B' C' D'). The corrected image has its columns oriented toward the north, with rows running east-west. The sub-satellite track is the ground trace of the platform carrying the scanner that collected the raw image. The line PP' of the raw image has the pixel centers indicated by squares, whereas the line QQ' of the corrected image has the pixel centers indicated by circles.

coordinates are referred to the longitude and latitude of the object. Usually, the image coordinates are known and the map coordinates are unknown. However, if the map coordinates are available for some reference points, a mapping function can be established between (x_n, y_n) and (X_n, Y_n), and applied to the remaining points, that is, in general:

$$\begin{cases} \hat{X}_n = f(x_n, \, y_n); \\ \hat{Y}_n = g(x_n, \, y_n). \end{cases} \tag{4.1}$$

By considering the 2-D polynomials, these functions are well known and applied to the image registration since the 1970s [267]. In the simplest case, that is, first-order polynomials, we have [103]:

$$\begin{cases} \hat{X}_n = a_{X0} + a_{X1}\, x_n + a_{X2}\, y_n; \\ \hat{Y}_n = a_{Y0} + a_{Y1}\, x_n + a_{Y2}\, y_n, \end{cases} \tag{4.2}$$

where the coefficients $a_{X0}, a_{X1}, a_{X2}, a_{Y0}, a_{Y1}$ and a_{Y2} have been obtained by applying a least squares algorithm on the available control points [181]. The optimization of the coefficients can be made globally (i.e., a single equation for the entire image) or locally (i.e., several equations, one for each image segment). Local optimization can be more accurate, but they usually require a higher computational effort [127].

A first-order function (six unknowns), or *affine transformation*, is suitable for the correction of rigid body distortions [51]:

$$\begin{cases} \hat{X}_n = t_x + s(x_n \cos\theta - y_n \sin\theta); \\ \hat{Y}_n = t_y + s(x_n \sin\theta + y_n \cos\theta), \end{cases} \tag{4.3}$$

where t_x and t_y denote translations in x and y axes, θ is the rotation angle, and $s(\cdot)$ is the scaling function. Even reflection distortions [131] can be compensated, but not warping effects, responsible for panoramic distortions and similar impairments. In this case, second-order polynomials (12 unknowns) or higher are necessary, even if in practice, polynomials of an order higher than three (20 unknowns) are rarely used. Actually, the literature has demonstrated that third-order polynomials introduce errors in relative pixel positioning in orthoimages, such as Landsat-TM or SPOT-HRV [56], as well as in geocoding and fusing multisensor images, such as SPOT-HRV and airborne SAR [248].

The 2-D polynomials are not able to correct the sources of distortions arising from the image formation (sensor nonlinearities, features observed from different viewpoints) and do not take into account the terrain relief effects, so that they can be used mainly for images with limited distortions, such as nadir viewing images, systematically corrected data by image providers, and small images over flat terrain [44]. Moreover, these functions are suitable for local distortions at the ground control point (GPC) locations, so that they are very sensitive to input errors and many and uniformly distributed GPCs are necessary [83]. As a consequence, 2-D polynomials are not suitable for multisource data integration requiring precise geometric positions and/or having high relief areas [249].

4.2.3.2   3-D polynomials

3-D polynomials are an extension of the 2-D case by adding a z-term related to third dimension of the terrain. However, even if the terrain term is taken into account, their limitations are similar to those of 2-D polynomials: they are very sensitive to input errors, they are applicable to small images with numerous and regularly distributed GCPs, and they are poorly robust and consistent in the operational environment. Better results can be obtained by using 3-D rational functions [249]. To overcome the problem of local geometric distortions, more specific mapping functions have been proposed, the most accurate of which is based on surface spline functions, which is however computationally complex and requires the design of faster local adaptive alternatives [101].

In general, procedures based on least squares functions are among the most used for referencing medium resolution images, as Landsat ETM+, which has a nominal spatial resolution of 30 m, even if the accuracy of the results depends on the number and distribution of GCPs. However, as stated before, if the optimization is global, the map and the image can be locally different, even in the control points. Moreover, empirical polynomials are very sensitive to the terrain relief, so that this problem can become critical for high-resolution images as the products of IKONOS and QuickBird sensors. To overcome this problem, GeoEye, the company owner of the IKONOS system, provides a set of rational polynomials with their stereo products [181].

4.2.4 Resampling

Once the coefficients of the transformation relating the coordinates of an image and its reference (map) have been computed, we can derive the equations relating the coordinates of the map and the image by inverting the transformation in such a way that (4.1) can be applied.

This is essential in order to define the values of the pixels located on the map sampling grid. In fact, the positions of the grid points on the map can be remapped onto the image grid and the pixel values at the corresponding coordinates can be assigned to the corresponding points of the map grid. This is conceptually simple but some practical trick is to be introduced to make the procedure feasible. In fact, image coordinates mapped from the map grid, in general, will not be integer values for which a pixel value can be directly identified on the image grid. More likely, such coordinates will be real values that identify a point in the neighborhood defined by four pixels of the image grid. In order to attribute a value to such a point and transfer this value on the map grid, an interpolation procedure is to be identified.

In practice, three modalities are commonly adopted. The first is denoted as the *nearest neighbor*. The value of the pixel of the image grid to be assigned to the map grid is chosen that is the nearest to the point of real coordinates remapped from the map grid.

The second is identified as *bilinear interpolation*. The weighted average of the four pixels of the image grid that are the nearest to the point mapped from the map grid is chosen as the pixel value of the map grid. The weights are proportional to the distance of the mapped grid points from the four image grid points. This is equivalent to an interpolate by a linear function both in x and y.

The last method commonly used for estimating pixel values in the map grid is *bicubic interpolation*. It is based on the fitting of two third-degree polynomials to the region surrounding the image point remapped from the map grid. In practice, the 16 nearest pixel values in the image are used to estimate the value of the pixel in the map image. This technique involves more computations than either the nearest neighbor or the bilinear methods with the advantage it gives more natural images without the blocking appearance of the nearest neighbor or the oversmoothing of the bilinear interpolation.

4.2.5   Other Registration Techniques

The point mapping methods described in Section 4.2.3 are essentially feature-based methods, because feature extraction and feature matching are involved. Feature-based algorithms can be performed on the spatial domain, by searching closed boundary regions, strong edges or very visible object as CPs [102, 164, 84], or on a transform domain, especially by considering multiresolution wavelet transforms and by extracting the desired features from the wavelet coefficients [160]. Another family of registration methods is constituted by area-based algorithms, in which the CPs are achieved by finding the subimage in the image that best matches a subimage in the reference image [103]. The comparison between subimages can be made by using several similarity metrics, as normalized cross correlation, correlation coefficient, mutual information, and other distances [139]. An exhaustive review of registration methods especially devoted to more recent techniques can be found in Le Moigne *et al.* [161].

4.3   Image Interpolation

Although seldom addressed in the literature on pansharpening, a crucial point of the protocol introduced in Section 4.1 is the interpolation of the MS image, in which spatial details coming from the Pan shall be injected, to the scale of the Pan image itself [151]. Actually, besides to original misregistrations that have not been well compensated, a theoretically inadequate interpolation procedure may induce additional subpixel shifts between the interpolated MS bands and the Pan image. In particular, pansharpening methods based on multiresolution analysis (MRA) [278, 196, 217, 12, 216, 202, 200], are intrinsically more sensitive to all types of misalignments between MS and Pan, than the complementary class of methods based on component substitution (CS) [45].

Interpolation is an essential part for most pansharpening techniques in order to bring the MS bands to the scale of the Pan image before the injection of Pan details into the MS image is performed. In the next sections, the theoretical fundamentals of digital interpolation are reported. Afterward, the reconstruction by means of local polynomial is considered and both odd and even one-dimensional interpolation kernels are formalized. Eventually, interpolation in the context of pansharpening is considered by reporting results of tests carried out on simulated Pléiades data.

4.3.1 Problem Statement

Theoretically, an image interpolation can be viewed to be composed by:

- a reconstruction process, either interpolation or approximation[1] [100], in which the underlying continuous intensity surface function is generated, starting from the original discrete data;

- a resampling process of the continuous intensity surface, so that a new digital image is produced, with a sampling frequency that is increased with respect to that of the original data.

In practice, however, the images have to be processed on digital devices, and, consequently, continuous functions cannot be addressed. In this case, therefore, a unique procedure that merges the two previous steps is necessary, so that the points of the resampled image can be obtained directly by means of a digital process. In most cases of interest for pansharpening (interpolation by integer scale factors and output sample grid displaced from the input one by an integer number of half pixels), interpolation can be performed by means of digital finite impulse response (FIR) filtering. This operation is done by using a discrete lowpass kernel, representing the sampled impulse response function, which is linearly convolved with an upsampled, that is, zero-interleaved, version of the original sequence of samples to yield the interpolated image at the desired scale [188].

4.3.2 Theoretical Fundamentals of Digital Interpolation

In order to develop the theory of digital interpolation, we briefly recall the theory of analogue signals interpolated by continuous-time filters.

4.3.2.1 Interpolation by continuous-time kernels

Let us consider the case of an input sequence $x[n]$ obtained by sampling a continuous-time intensity function $x_C(t)$, that is, $x[n] = x_C(nT_x)$, so that the conditions of the sampling theorem are fulfilled, that is, $x_C(t)$ is bandlimited with frequency spectrum $X_C(j)$ that is zero for $|\Omega| > \Omega_N$, and sampled at a rate $\Omega_x > 2\Omega_N$, where $\Omega_x = 2\pi/T_x$, being T_x the sampling period of

---

[1]The term *interpolation* indicates that the input sample sets are samples of the continuous function, whose resampling yields the output sample set, unlike *approximation*, in which the continuous function does not pass through the values of the input samples.

the sequence $x[n]$. The continuous function $x_C(t)$ can be recovered from the original discrete sequence by convolving a continuous-time normalized impulse train $x_S(t)$ of period T_x and modulated by the samples $x[n]$,

$$x_S(t) = T_x \cdot \sum_{n=-\infty}^{+\infty} x[n]\,\delta(t - nT_x), \qquad (4.4)$$

with the impulse response $h_R(t)$ of a lowpass continuous-time reconstruction kernel having a frequency response $H_R(j\Omega)$:

$$x_C(t) = x_S(t) * h_R(t) = T_x \cdot \sum_{n=-\infty}^{+\infty} x[n]\,h_R(t - nT_x), \qquad (4.5)$$

where the symbol $*$ denotes a linear convolution [199]. Normalization of the impulse train ensures that $x_S(t)$ and $x_C(t)$ have the same dimension of waveforms.

In the frequency domain, the base period, $X_C(j\Omega)$, is selected in the periodic Fourier transform $X(e^{j\omega}) = X(e^{j\Omega T_x})$ of the input sequence $x[n]$, by removing all the scaled replicas of $X_C(j\Omega)$ located at integer multiples of the sampling frequency Ω_x, or, equivalently, at integer multiples of 2π, in the normalized frequency domain $\omega = \Omega T_x$, which is the standard domain for representing the spectra of digital sequences.

The reconstructed continuous-time function $x_C(t)$ can be now resampled at a higher sampling frequency Ω_y, with $\Omega_y = L\Omega_x$, and a related sampling period of $T_y = T_x/L$, being $L > 1$ the interpolation factor, so that an interpolated sequence $y[m]$ having a sampling rate L times greater than that of the original sequence $x[n]$ is generated. The normalized frequency spectrum of $y[m]$, named $Y(e^{j\omega})$, will be a scaled periodic repetition of $X_C(j\Omega)$ placed at integer multiples of 2π on the normalized frequency $\omega = \Omega T_y$ axis [199].

4.3.2.2 Interpolation by digital kernels

The interpolated sequence $y[m]$ can be equivalently obtained by means of discrete operations only. Figure 4.4 shows the block diagram of a digital interpolation by L of the input sequence $x[n]$.

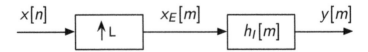

FIGURE 4.4: Flowchart of the digital interpolation by a factor L of the input sequence $x[n]$.

First, a discrete-time sequence $x_E[m]$ is generated by inserting $L-1$ null samples between any two consecutive samples in the original sequence $x[n]$, so that $x[n]$ is expanded at the size of the interpolated sequence $y[m]$. The normalized frequency spectrum $X_E(e^{j\omega})$ of $x_E[m]$ is a periodic repetition of $X_C(j\Omega)$, scaled in amplitude and possibly rotated in phase, and located at integer multiples of $2\pi/L$. The recovery of the normalized frequency spectrum of $y[m]$ can be done by means of a digital lowpass kernel $h_I[m]$, which has to be able to remove all the spectral replicas except those at integer multiples of 2π. Consequently, the normalized frequency response $H_I(e^{j\omega})$ of $h_I[m]$ is a 2π scaled repetition of the frequency response of the continuous-time lowpass filter $H_R(j\Omega)$. The impulse response $h_I[m]$ can be thus obtained by sampling the continuous-time impulse response $h_R(t)$ with a sampling period of $T_y = 1/L$ [199].

Figure 4.5 depicts the time and frequency domain representations of the original sequence, $x[n]$, the upsampled sequence, $x_E[m] = x[n] \uparrow L$, and the interpolated sequence $y[m] = x_E[m] * h_I[m]$, in the case of $L = 3$. The frequency spectra are $X(e^{j\Omega T_x})$, $X(e^{j\Omega T_x/3})$, and $X(e^{j\Omega T_x/3}) \cdot H_I(e^{j\Omega T_x/3})$, respectively. Note that in Figure 4.5(b), the frequency axes are not normalized to the respective sampling frequency intervals, to avoid the presence of different scales in the time domain plots of Figure 4.5(a).

Even if the filtering block that implements the impulse response $h_I[m]$ is linear and time invariant (LTI), the entire system (upsampler followed by lowpass filter) is characterized by an impulse response $h_E[n, m]$ that is linear time variant, that is, if the input is shifted an integer number of samples δ, the output is not shifted of the same quantity, unless δ is a multiple of L. In particular, $h_E[n, m]$ is periodic in m with period L, so that the interpolator by L can be viewed as a linear periodically time varying discrete-time system. A polyphase structure can be used for representing the I/O relationship between $x[n]$ and $y[m]$, so that the sequence $y[m]$ is regarded as the interleaving of L subsequences of length m/L. The computation of each output subsequence starting from the input sequence $x[n]$ involves a different filter, though all filters are derived from $h_I[m]$. In this way, multiplications by zero samples are implicitly disregarded and computation is faster, especially for large L [76, 186].

4.3.3 Ideal and Practical Interpolators

A large variety of interpolating functions have been proposed in the literature for an optimal image reconstruction [177, 188, 207, 223, 233]. From a theoretical point of view, the points on the continuous surface can be exactly recovered starting from the sampled values, once the sampling frequency has been taken above the Nyquist rate, that is, twice the signal bandwidth, if the original signal is bandlimited; in this case, the interpolating function is an ideal lowpass filter, that is, a *sinc* function.

Ideally, such a procedure would require images of infinite extent and an interpolation filter having infinite length. However, the original image has finite extent and a practical reconstructor should be fast, especially if large amounts of data are to be processed, as in image pansharpening. Therefore,

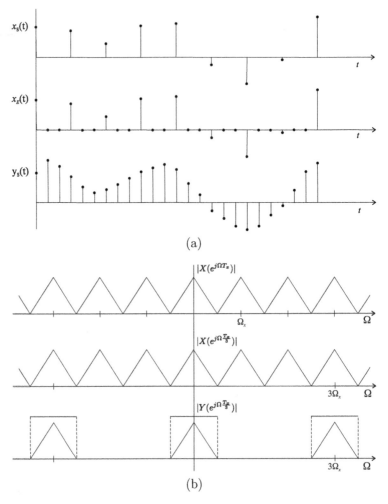

FIGURE 4.5: Digital interpolation by a factor $L = 3$: (a) time domain representation; (b) frequency domain representation.

local interpolators, which involve only few samples in a limited neighborhood of the point to be reconstructed are commonly adopted [62]. Among them, piecewise local polynomials are characterized by a very good trade-off between computational complexity and quality of the reconstructed image [206, 233]. Conversely, theoretically optimal *sinc* interpolating functions need a windowing, that is, a smooth truncation, of their infinite impulse response, to avoid Gibbs ringing impairments. Consequently, the design of finite duration interpolators, such as those achieved from a finite length polynomial approximation of the *sinc* function, is preferable to the windowing of the *sinc*, especially for the total absence of ripple of the former [12].

4.3.4 Piecewise Local Polynomial Kernels

Image interpolation by means of piecewise local polynomials has been extensively used in the literature, especially with nearest neighbor (NN), linear and cubic interpolators. NN (degree 0), linear (degree 1), and cubic (degree 3) interpolators are widespread for pansharpening and available in image processing software packages. However, the NN kernel, which performs a replication of input samples, is practically used only for strictly real-time applications, which are satisfied even by low quality products, because of its poorly selective frequency response, which introduce evident blocking impairments. Concerning the quadratic (degree 2) interpolator, it has been recognized in the past to be unsuitable for image upsampling because of a nonlinear phase, which seemed to introduce distortions [223, 266]. Instead, such a conclusion appeared later to be caused by inappropriate assumptions about the filter design [88, 89], so that a correct implementation does not introduce any phase distortion and exhibits performances comparable to those of the cubic kernel.

An important difference between even order and odd order interpolators is that NN and quadratic interpolators produce rigid half pixel shifts in the upsampled images because of their even symmetry, and hence linear phase, if they are used for interpolations by even factors [207]. Conversely, linear and cubic interpolators can be implemented as zero phase filters of odd length, regardless of the interpolation factor, so that the produced images do not have any shifts with respect to the originals. However, the apparent drawback of the even order filters may become an important asset in the case of image pansharpening, because of the acquisition characteristics of modern MS and Pan scanners, whose point spread functions (PSF) are centered in the middle of each pixel. The remainder of this chapter will show that the direct consequence of this misalignment is the rising of a shift to be compensated between Pan and MS images. This shift is equal to an integer number of pixels if the scale ratio is odd, while it is an integer number plus a half of pixel if the scale ratio is even, as in most cases of interest. It should be noted that the scale ratio between Pan and MS is obviously equal to the interpolation factor to be applied to the MS images. Therefore, if an odd scale ratio is given, an odd filter can be utilized. In this case, polynomials of all degrees are feasible. Conversely, if, as usual, an even factor is present between the two scales, NN and quadratic interpolators are intrinsically capable of producing superimposed MS and Pan data. It is noteworthy, however, that also linear and cubic filters can be designed as even filters and adopted for MS interpolation.

4.3.5   Derivation of Piecewise Local Polynomial Kernels

Digital interpolation can be effectively performed by piecewise local polynomial kernels. In order to obtain the coefficients of a kernel, we have to first consider the correspondent continuous-time kernel and afterward sample it in even or odd modality as shown in detail in the next sections.

4.3.5.1   Continuous-time local polynomial kernels

In Section 4.3.2, we have shown that the interpolated sequence $y[m]$ is obtained by filtering the upsampled sequence $x_E[m]$ by means of a digital lowpass filter whose impulse response $h_I[m]$ is generated by sampling the continuous-time impulse response $h_R(t)$, with a sampling period of $T_y = T_x/L$.

The design of the continuous-time filter $h_R(t)$ can be made as follows. Since the ideal interpolating *sinc* function has infinite length and thus cannot be implemented as an FIR filter, let us focus on polynomial approximations of *sinc*, as specified in Section 4.3.4. In this case, the interpolating curve will be constituted by an ensemble of one sample wide polynomial pieces (piecewise local interpolation).

If p denotes the degree of each polynomial piece, each piece is a function of the $p+1$ samples of the original sequence that are closest to the sample to be interpolated. So, if $p+1$ is even, then the degree of the polynomial piece is odd (linear and cubic reconstructors), and each piece starts and ends at adjacent sample points. Instead, if $p+1$ is odd, the degree of the polynomial piece is even (zero-order and quadratic reconstructors), and each piece starts and ends halfway between two adjacent sample points [88].

TABLE 4.2: Continuous-time impulse responses of piecewise local polynomial kernels.

Polynomial Degree	Continuous-Time Impulse Response $h_R(t)$	
$p=0$ (NN)	1 0	if $\lvert t \rvert \leq \frac{1}{2}$ otherwise
$p=1$ (linear)	$1 - \lvert t \rvert$ 0	if $\lvert t \rvert \leq 1$ otherwise
$p=2$ (quadratic)	$-2\lvert t \rvert^2 + 1$ $\lvert t \rvert^2 - \frac{5}{2}\lvert t \rvert + \frac{3}{2}$ 0	if $\lvert t \rvert \leq \frac{1}{2}$ else if $\frac{1}{2} < \lvert t \rvert \leq \frac{3}{2}$ otherwise
$p=3$ (cubic)	$\frac{3}{2}\lvert t \rvert^3 - \frac{5}{2}\lvert t \rvert^2 + 1$ $-\frac{1}{2}\lvert t \rvert^3 + \frac{5}{2}\lvert t \rvert^2 - 4\lvert t \rvert + 2$ 0	if $\lvert t \rvert \leq 1$ else if $1 < \lvert t \rvert \leq 2$ otherwise

Table 4.2 shows examples of reconstruction kernels $h_R(t)$ for zero-order, linear, quadratic, and cubic reconstructors in the case of unity-spaced samples

[89]. The assumption $T_x = 1$ will hold hereinafter for simplicity and ease of notation, without loss of generality. Each value of the continuous function $x_C(t)$ is recovered by superimposing the interpolating polynomial $h_R(t)$ to the underlying point and multiplying each available value $x_C(n)$, $t - \frac{p+1}{2} \le n \le t + \frac{p+1}{2}$ by the corresponding value $h_R(t - n)$.

4.3.5.2   Odd and even local polynomial kernels

Let us now consider the correspondent digital kernels having an impulse response $h_I[m]$ that is obtained by sampling the continuous-time impulse response $h_R(t)$. One of the most important design parameters is the length of $h_I[m]$, that is, the number M of its coefficients. If $M = 2\tau + 1$, with τ not necessarily integer, then the filter will have a linear phase with a delay of τ samples *if and only if* the impulse response is symmetric, that is, $h_I[m] = h_I[M - 1 - m]$ [199]. This constraint can be fulfilled by two distinct impulse response shapes, denoting an *even* or an *odd* filter, respectively, and depending on whether M is odd or even, as shown in Figure 4.6.

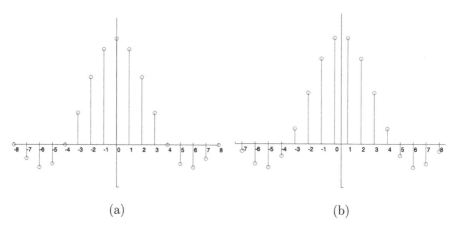

(a)   (b)

FIGURE 4.6: The two possibilities for symmetric impulse responses in linear phase FIR filters for 1:4 interpolation: (a) M odd; (b) M even. For better readability, the origin is placed in $(M - 1)/2$ and $M/2 - 1$ for M odd and M even, respectively, that is, filters are noncausal, and time axes are shifted accordingly.

If M is even, the symmetry center lies midway between the two middle samples $M/2 - 1$ and $M/2$ and the delay $\tau = (M - 1)/2$ is not an integer. All samples in the expanded sequence lie in positions not previously occupied by original samples. Conversely, if M is odd, the symmetry center is in the middle sample $(M - 1)/2$, and the delay is an integer. This filter is suitable if the samples of the original sequence have to be retained in the interpolated sequence.

Since the spacing between two consecutive samples of the impulse response is constrained to $1/L$, an increment in M implies that more points of the original sequence are involved in the computation of the interpolated value, approximately M/L samples. By considering piecewise polynomials, the number of pixels involved in the computation is $q = p + 1$, where p is the degree of each polynomial piece. The length of the polynomial kernel will be $M = q(L - 1) + 2\lfloor \frac{q+1}{2} \rfloor - 1$ in the case of an odd kernel, and $M = qL$ in the case of an even kernel. In the case of an odd kernel, the value of M has to be odd. Therefore, if L is odd, then $L - 1$ is even and M is certainly odd. The consequence is that an odd kernel is implementable for each value of q, that is, for all degrees of the polynomial function. Concerning the case of an even kernel, the value of M has to be even as well. Consequently, the value of q is necessarily even, and only linear or cubic kernels are implementable. In any case, if L is odd, even filters are not necessary to match the necessities of satellite sensors.

Conversely, if L is even, then $L - 1$ is odd and the value of M is odd, that is, an odd kernel is available, only if q is even, for example, in the case of linear or cubic kernels. If q is odd, for example, in the case of NN or quadratic kernels, only an even filter is implementable and, consequently, a noninteger shift of a half a pixel plus a possible additional shift of an integer number of pixels will necessarily occur between the input and output samples. However, if L is even, even filters can be implemented also by means of linear and cubic polynomials because the value of M is forced to be even for all values of q.

Thus far, causal filters usually adopted in signal processing have been addressed. These filters are suitable mainly for time function signals because the filter can process only values already available for the user. However, when digital images are processed, noncausal filters are preferred, because the image is usually available on the whole and the time constraint is no longer needed. The difference is that causal filters having symmetric impulse response always exhibit a nonzero linear phase term, and hence a nonzero delay τ, which can be either integer or noninteger, depending on whether the filter length is odd or even, respectively. Concerning noncausal filters, they are zero phase filters if their discrete impulse response has an odd length; otherwise a nonzero linear phase term is present, and consequently a subpixel delay τ, which is half of pixel in the case of even filters. In the following, we will consider noncausal filters, so that odd filters have zero phase, and hence zero delay, and even filters have linear phase with a half of pixel delay.

Tables 4.3 and 4.4 report the coefficients of the digital LTI lowpass FIR filter $h_I[m]$ for even and odd polynomial kernels with interpolation factors $L = 2$ and $L = 3$. Note that the sum of the coefficients is always equal to the interpolation factor. The numerical values of coefficients indicate that implementations with finite arithmetics are preferable.

Figure 4.7 shows the normalized amplitude responses of the polynomial kernels of odd and even length, as reported in Table 4.3, for an interpolation factor of $L = 2$. The plots in logarithmic scale (magnitude) highlight that polynomial kernels exhibit no ripple in either passband or stopband. The

TABLE 4.3: Digital LTI FIR kernels $\{h_I[n]\}$, obtained by sampling the polynomials of Table 4.2, for $L = 2$.

Filter Type	Odd	Even
$p = 0$ (NN)	unfeasible	$\{1, 1\}$
$p = 1$ (linear)	$\{\frac{1}{2}, 1, \frac{1}{2}\}$	$\{\frac{1}{4}, \frac{3}{4}, \frac{3}{4}, \frac{1}{4}\}$
$p = 2$ (quadratic)	unfeasible	$\{-\frac{1}{16}, \frac{3}{16}, \frac{14}{16}, \frac{14}{16}, \frac{3}{16}, -\frac{1}{16}\}$
$p = 3$ (cubic)	$\{-\frac{1}{16}, 0, \frac{9}{16}, 1, \frac{9}{16}, 0, -\frac{1}{16}\}$	$\{-\frac{3}{128}, -\frac{9}{128}, \frac{29}{128}, \frac{111}{128}, \frac{111}{128}, \frac{29}{128}, -\frac{9}{128}, -\frac{3}{128}\}$

TABLE 4.4: Digital LTI FIR kernels $\{h_I[n]\}$, obtained by sampling the polynomials of Table 4.2, for $L = 3$.

Filter Type	Odd
$p = 0$ (NN)	$\{1, 1, 1\}$
$p = 1$ (linear)	$\{\frac{1}{3}, \frac{2}{3}, 1, \frac{2}{3}, \frac{1}{3}\}$
$p = 2$ (quadratic)	$\{-\frac{1}{18}, 0, \frac{5}{18}, \frac{14}{18}, 1, \frac{14}{18}, \frac{5}{18}, 0, -\frac{1}{18}\}$
$p = 3$ (cubic)	$\{-\frac{2}{54}, -\frac{4}{54}, 0, \frac{18}{54}, \frac{42}{54}, 1, \frac{42}{54}, \frac{18}{54}, 0, -\frac{4}{54}, -\frac{2}{54}\}$

Filter Type	Even
$p = 0$ (NN)	unfeasible
$p = 1$ (linear)	$\{\frac{1}{6}, \frac{3}{6}, \frac{5}{6}, \frac{5}{6}, \frac{3}{6}, \frac{1}{6}\}$
$p = 2$ (quadratic)	unfeasible
$p = 3$ (cubic)	$\{-\frac{35}{1296}, -\frac{81}{1296}, -\frac{55}{1296}, \frac{231}{1296}, \frac{729}{1296}, \frac{1155}{1296}, \frac{1155}{1296}, \frac{729}{1296}, \frac{231}{1296}, -\frac{55}{1296}, -\frac{81}{1296}, -\frac{35}{1296}\}$

plots in Figure 4.7(a) and (b) are relative to polynomial orders up to the 11th and are taken from Aiazzi *et al.* [12]. The purpose of the lowpass filter for interpolation by two is to completely reject the upper half of the band defined by the Nyquist frequency (0.5 in normalized plots), while thoroughly retaining the lower half band. Thus, the ideal interpolator is an infinite length *sinc*, whose frequency response is the rectangle with a dashed vertical edge superimposed to all plots.

Therefore, the more similar is the frequency response of a filter, both in passband and in stopband, to the ideal rectangle, the better is its performance for interpolation. In this sense, Figure 4.7(a) reveals that better and better zero-phase filters can be obtained by increasing the order of the polynomial. However, the most relevant advantage occurs when the linear interpolator ($p = 1$) is replaced with a cubic one ($p = 3$). Figure 4.7(c) presents both interpolators of even degrees, admitting only implementations of even lengths, and even implementations of interpolators of odd degrees. What immediately stands out is that the frequency responses of even filters do not pass through the point (0.25, 0.5), that is the center of the shaded cross, but their distance

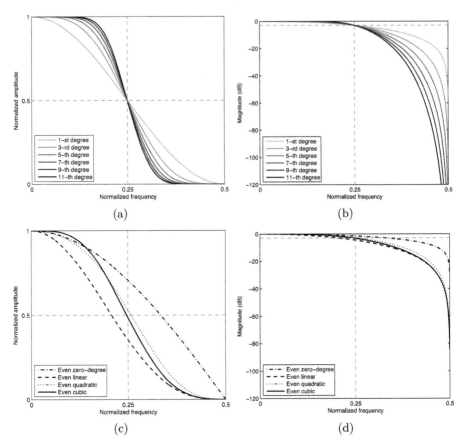

FIGURE 4.7: Frequency responses of piecewise local polynomial interpolating kernels: (a): zero-phase symmetric kernels of odd length for interpolation of degrees 1, 3, 5, 7, 9, and 11 (linear scale); (b): logarithmic scale; (c): linear phase symmetric kernels of even length for interpolation of degrees 0, 1, 2, and 3 (linear scale); (d): logarithmic scale.

to that point tends to be zero, as the order of the polynomial increases. The response of the NN interpolator (zero-degree polynomial) is half of the main lobe of a *sinc*: it is adequate in the passband, but little selective in stopband. Conversely, the linear interpolator of even length exhibits an adequately selective response in the stopband, but a poor response in the passband. Quadratic and even cubic interpolators are much better and quite similar to each other, the latter being slightly better, especially in the stopband.

Figure 4.8 compares the frequency responses of odd and even implementations of linear and cubic interpolators. It is noteworthy that while the linear interpolator of even length is somewhat poorer than its odd length version (amplitude lower than 0.5 at the cutoff frequency, scarce preservation in the

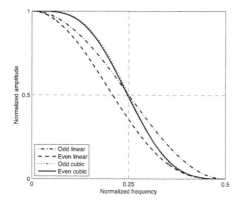

FIGURE 4.8: Frequency responses of piecewise local polynomial interpolating kernels: a comparison of odd and even length implementations of linear and cubic interpolators.

passband), the odd length and even length implementations of the cubic interpolator are practically identical to each other, the former being imperceptibly better. In substance, their frequency responses only differ by the phase term, either zero or linear with a half sample shift. Therefore, excellent performance of even cubic interpolators can be achieved in application contexts in which even interpolation is preferable, as pansharpening is.

4.3.6 Interpolation of MS Data for Pansharpening

In this section, the acquisition geometry of very high-resolution imaging instruments is presented and discussed in order to show how even and odd interpolators act when the MS bands have to be upsampled to the Pan resolution and overlap the Pan without geometric misregistration. First, the one-dimensional case is reported and after it is extended to 2-D images.

4.3.6.1 One-dimensional case

Let us consider first the one-dimensional case for simplicity of notation. Let $P(x)$ be the (unknown) intensity surface imaged by the Pan sensor and $MS(x)$ the (unknown) intensity surface imaged by any of the channels of the MS instrument. Let $P[n_p] = P(n_p T_p)$ denote the Pan image sampled with a sampling step size of T_p. Let also $MS[n_m] = MS(n_m T_m)$ denote the sampled MS image, having a sampling step size of T_m. In pansharpening data fusion, T_m is an integer multiple of T_p, that is, $T_m = r T_p$, r being the scale factor between MS and Pan, from which $n_p T_p = n_m T_m$, and thus $n_p = r n_m$.

Let us suppose that the Pan and MS sensors image the same scene, starting from $x = 0$, and ending to $x = X - 1$. By assuming a unity sampling step size for the Pan, then the sample points of the Pan sensor are located at the coordinates $x_i = i$, $i = 0, 1, \ldots, X - 1$. If the PSFs of the sensors were centered on the left side of each pixel, the sample points of MS would be aligned to those of Pan and located at the coordinates $x_i = r \cdot i, i = 0, 1, \ldots, \frac{X}{r} - 1$. In this case, all the values of the original MS images could be recovered in the interpolated MS bands at the scale of Pan. For each point of the Pan image, only $r - 1$ points of the MS bands should be computed, and an odd filter would be ideal for interpolation.

However, the actual situation in satellite instruments is that all PSFs are centered in the middle of corresponding pixels. The consequence is that the coordinates of the sample points of the MS sensor become:

$$x_i = \frac{r - 1}{2} + r \cdot i, \qquad i = 0, 1, \ldots, \frac{X}{r} - 1. \tag{4.6}$$

If the scale factor r is odd, then the delay τ is an integer shift of the Pan coordinates that can be easily compensated, without resampling the interpolated image: a filter of odd length is still preferable. However, if r is even, as it usually happens ($r = 4$ for the majority of instruments), then the delay τ is noninteger, that is, interpolated samples lie in the middle of an integer shift of the Pan coordinates. In order to obtain perfectly superimposed MS and Pan, the values of the interpolated MS bands have to be computed by means of a filter of even length.

4.3.6.2 Two-dimensional case

The concepts expressed for the one-dimensional case are now extended to the two-dimensional one. Figure 4.9 shows the acquisition geometry of actual MS and Pan instruments, for the two cases of scale ratios between the MS and Pan images equal to an even number (two and four) and to an odd number (three). In the former case, MS pixels lie in the middle of either 2×2 or 4×4 blocks of Pan pixel. Consequently, the grid of MS and Pan pixels is intrinsically shifted by half the size of a Pan pixel along both rows and columns.

Interpolation by means of an odd filter will produce a grid of interpolated MS pixels having same spacing as the grid of Pan, but still being displaced by half a pixel. In the latter case, an MS pixel lies in the middle of a 3×3 block of Pan pixels. Thus, interpolation will produce a grid of MS pixels that is shifted by one pixel to the grid of Pan pixels, according to (4.6).

For the sake of clarity, let us recall how an odd and an even filter interpolate the MS image. Interpolation by four is performed separably in two steps of interpolation by two. Figure 4.10(a) shows the initial 2×2 set of MS pixels, the intermediate set of MS pixels interpolated by two and the final set of MS pixels interpolated by four, as achieved by an interpolating filter of odd length. As it appears, the final grid of samples contains both the original samples and those generated in the first stage of interpolation by two.

Instead, Figure 4.10(b) shows the initial, intermediate, and final MS pixels, as achieved by an interpolator of even length. Unlike odd interpolation, the input and output grids are always half-pixel displaced and each stage generates

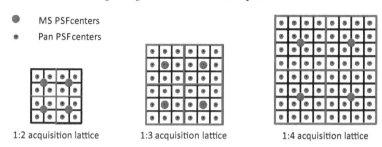

FIGURE 4.9: Acquisition geometry of an MS and a Pan scanner for scale ratios 1:2, 1:3, and 1:4 between Pan and MS pixel sizes.

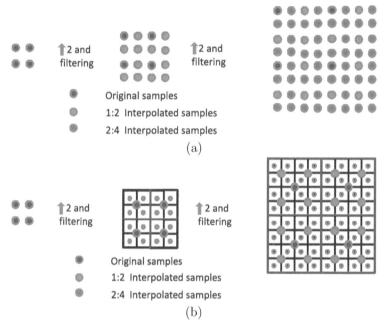

FIGURE 4.10: Two-dimensional interpolation by four achieved as two cascaded steps of interpolation by two with: (a) odd length filter; (b) even length filter. Note that in interpolation with an odd filter, initial and intermediate samples belong to the final interpolated sample set; but not with an even filter.

a grid of pixels that is displaced by half a pixel from the input grid and thus does not contain the input pixels. If we compare the final situation with that in Figure 4.9, we shall note that the grid of MS pixels interpolated at the scale of Pan is perfectly superimposed to the grid of Pan pixels generated by the instrument. Instead, if an interpolation by four is achieved by means of an odd filter, the output grid of MS pixels is shifted from the grid of Pan pixels by one pixel and one-half along both rows and columns. The integer part of such shifts can be compensated by means of a simple rigid translation of the interpolated image. However, the fractional part can be compensated only through a resampling, which implies another interpolation, this time to a sampling grid of same spacing but shifted by half a pixel. Note that if the interpolation by four is performed by means of a unique step, then only the fractional part of the shift is present.

4.3.7 Interpolation Assessment for Pansharpening

The assessment of the interpolation procedures has been exploited by considering how a correct or incorrect interpolation affects pansharpening results. After the presentation of the dataset that has been chosen for the simulations, an extensive series of results is presented by discussing the main representative cases of interest. In particular, odd and even interpolators of different orders have been applied to images acquired with and without a relative shift.

4.3.7.1 Evaluation criteria

Quality assessments have been carried out according to the synthesis property of Wald's protocol (see Section 3.2.2.1). Image-to-image *vector* indices, like SAM (3.8), ERGAS (3.10), and Q4 (3.12) have been calculated for each fusion simulation.

4.3.7.2 Dataset

From 1999 to 2002, CNES (Centre National d'Études Spatiales–French Space Agency) led a research program aimed at identifying and evaluating MS pansharpening fusion method on inframetric spatial resolution images. For this study, images have been acquired from an airborne platform over nine sites covering various landscapes (rural, urban, or coastal). Five methods, selected as the most efficient, were compared on this dataset. The evaluation was both qualitative with visual expert opinions and quantitative with statistical criteria [158]. The simulated Pléiades dataset was also used for the IEEE Data Fusion Committee (DFC) Contest [37].

Starting from four MS bands collected by an aerial platform at 60 cm resolution, the following products were synthesized by CNES:

- High-resolution Pan image at 80 cm obtained by (a) averaging the green and red channels, (b) applying the nominal MTF of the Pan scanner (amplitude about 10% Nyquist frequency), (c) resampling the outcome to 80 cm, (d) adding the instrument noise, (e) recovering the ideal image by means of inverse filtering and wavelet denoising.

- Four MS bands with spatial resolution equal to that of Pan obtained by applying the nominal MTFs of each spectral channel of the spaceborne instrument, implemented as nonseparable zero-phase FIR filters, and resampling the outcome to 80 cm.

- Four MS bands at 3.2 m (resolution four times lower than that of Pan) simulated from the 80 cm MS product, again by applying the MTFs of each spectral channel and decimating the outcome by four along rows and columns.

Since the Pan image is synthesized from G and R channels and all zero-phase FIR filters have been used, the 80 cm MS image is perfectly registered to the Pan image and the 3.2 m MS image exhibits what we will refer to as an *odd alignment* with the 80 cm Pan. This image does not reflect the typical situation depicted in Figure 4.9, which can be referred to as an *even alignment*. In order to generate 3.2 m MS data with *even alignment*, the spectral PSFs of Pléiades, available as zero-phase nonseparable FIR filters, were transformed to linear phase filters, by resampling the continuous PSFs reconstructed from the coefficients of the original FIR filters with half-pixel shifts (see Figure 4.6).

Eventually, 80 cm Pan, 80 cm reference MS originals and two 3.2 m MS versions, one with odd alignment and another with even alignment, are used for fusion simulations and quality assessments. The size of the simulated Pan image is approximately 5000×20000 pixels; the four *reduced* MS bands are 1250×5000. Only the scene portraying the city of Toulouse is considered in this chapter.

The main advantage of using simulated data is that fusion performances can be objectively evaluated at the same spatial scale of the final user's product, unlike in fusion assessments carried out on spatially degraded versions of commercial data, in which interpolation will depend on reduction filters used before decimation to prevent aliasing. Thus, the problem of interpolation cannot be decoupled from that of decimation, unless simulated data having references at a full resolution are employed, as in the present case.

4.3.7.3 Results and discussion

The reported fusion simulations have been carried out on the simulated Pléiades data, both on the original version with 3.2 m MS odd-aligned to 80 cm Pan used in Laporterie-Déjean *et al.* [158] and Alparone *et al.* [37], and on the purposely generated version with MS even aligned to Pan. Note that the former case is unrealistic and occurs because Pan data have been synthesized from high-resolution MS data, whereas the latter exactly matches the MS-to-Pan alignment featured by real spaceborne instruments (IKONOS, QuickBird, GeoEye, WorldView, etc.).

The goal of the simulations is to highlight the benefits of a correct interpolation, that is, an interpolation that does not introduce misalignment between expanded MS and Pan data. Moreover, the trend in performance indices (Q4, ERGAS, and SAM) with the order of interpolation is evidenced, or equivalently with the degree of the local interpolating polynomial. In the simulations, odd implementation of degrees 1, 3, 7, and 11 and even implementations of degrees 0 (NN interpolator), 1, 2, and 3 are considered. Note that even implementations of polynomials of odd degrees, like the cubic one, have only recently been addressed in the literature. Actually, odd implementations have already been developed for a long time and there has been no real motivation, apart from the acquisition geometry of MS-Pan instruments and perhaps the interpolation of 4:2:0 decimated chroma components for video codecs, to develop even ones.

The two types of interpolations have been tested by considering the differences in performances obtained by running a typical fusion method based on the "à-trous" wavelet transform (ATW). Figure 4.11 reports SAM, ERGAS, and Q4 for the ATW fusion method by considering different alignments of the input data set, that is, odd and even alignments between MS and Pan and odd and even implementation of interpolating polynomial filters. The performances have been reported varying with the degree of the polynomials. It is noteworthy that only odd filters with odd aligned data and even filters with even aligned data produce perfectly overlapped Pan and MS interpolated by four. Instead the other two combinations (even filters, odd data; odd filters, even data) introduce a diagonal shift by 1.5 pixels.[2] Misalignments are the main reasons for the poorer performances of one class of filters, odd or even, with respect to the other.

From the values of scores in Figure 4.11, some considerations can be made. For odd filters, whenever suitable, fusion performances steadily increase with the order of the interpolating polynomial. However, about two-thirds of the performance increment between the linear and 11th order interpolation is achieved by the cubic kernel, which deservedly represents the best trade-off. For even filters, the trend is not monotonous, at least for SAM and Q4: NN interpolation is preferable to the even linear interpolation, which yields the worst results. The frequency responses of the two filters, shown in Figure 4.7(c), evidence that preservation in passband is more important than rejection outside passband. That is also one reason for which NN interpolation is widely used for pansharpening. Among even kernels, the quadratic one, introduced in Dodgson [89] as a continuous-time filter, always outperforms NN and linear interpolators, but the even cubic kernel is always superior.

Differences in performances due to the misregistration appear evident for all the indices. By considering the spatial shifts of colors caused by MS to Pan misregistration, we can note that SAM is adequate for capturing changes in color hues, but is less powerful for detecting geometric changes. This effect is visible in Figure 4.12, where images before and after fusion, together with a high-resolution reference original are displayed. The scale has been deliberately set in order to appreciate small shifts (1.5 pixels). That is the reason by which high-resolution MS and Pan originals and all fusion prod-

---

[2]Separable interpolation by a factor L along both rows and columns carried out with symmetric filters of even lengths introduces shifts equal to $(L-1)/2$ along the same directions between input and output samples.

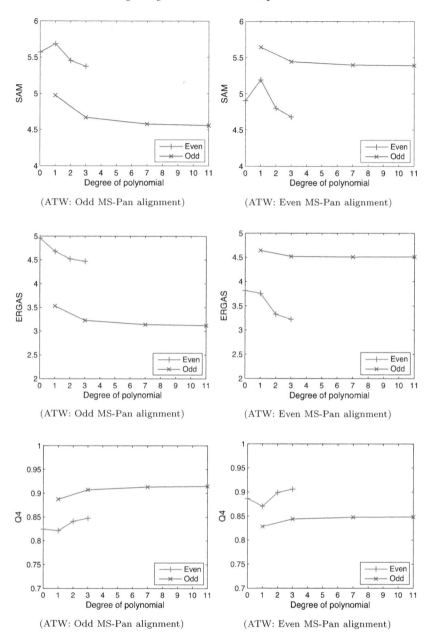

FIGURE 4.11: Quality (Q4) and distortion (SAM, ERGAS) indices of a wavelet-based fusion method (ATW) for different alignments of MS to Pan and for polynomial kernels of odd and even lengths, as a function of the degree of an interpolating polynomial.

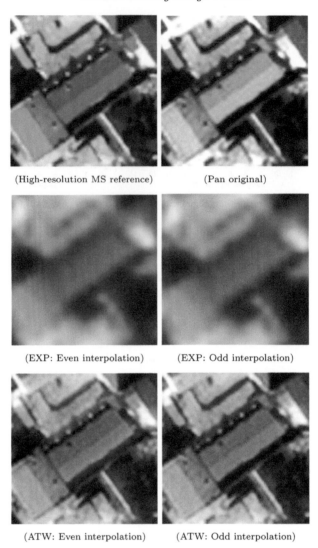

(High-resolution MS reference)   (Pan original)

(EXP: Even interpolation)   (EXP: Odd interpolation)

(ATW: Even interpolation)   (ATW: Odd interpolation)

FIGURE 4.12: Pan and true-color composition of MS details interpolated by means of even and odd length cubic filters, MS high-resolution reference and fusion products. The MS reference is exactly superimposed to Pan. Original MS data before interpolation are even-aligned to Pan, as in Figure 4.9, case 1:4. Interpolation with an even filter of original MS data yields no shifts between MS and Pan. Interpolation with an odd filter introduces an 1.5 pixel diagonal shift between MS and Pan. Fusion products of ATW starting from odd and even bicubically interpolated images show that the incorrect odd interpolator causes color shifts, mainly visible on the roof. The size of all icons is 64×64.

ucts are moderately sharp. The correct interpolation is with a filter of even length. Bicubic interpolation, either odd or even, has been used throughout. ATW fusion products are good when interpolation is even, but the method is somewhat poor when interpolation is odd, that is, not correct.

In the literature, several interpretations have been made concerning the most suitable interpolation technique to be implemented for pansharpening. Some authors generically suggest the use of either NN or bicubic interpolation. Others recommend the use of the bicubic one because it is more similar to the ideal interpolating *sinc*, which yields best performances. Others more claim that NN is preferable to bicubic because the former does not change the radiometric values of pixels, unlike the latter does. Actually, NN is preferable to bicubic interpolation when the data are even-aligned, because NN, being a filter of even length, compensates the 1.5 MS to Pan pixel shift occurring after interpolation; bicubic interpolation, at least in its standard odd version, does not. Figure 4.11 highlights that, whenever MS and Pan are even-aligned, NN is significantly preferable to bicubic odd interpolation, in terms of all scores (around 10% better in average). However, NN interpolation is about 5% poorer than the cubic interpolation of even length. However, such a difference depends only on the quality of interpolation, not on disregarded alignment errors.

As a summarizing example, let us suppose to interpolate by using a cubic kernel on even-aligned MS data, as it is the most common practical case. Let us consider again the ATW fusion method. Table 4.5 reports performance indices for the following four cases.

1. Even alignment is disregarded and a conventional zero-phase (odd) cubic kernel is used for interpolation.

2. Even alignment is partially compensated by means of a rigid diagonal shift by one pixel, so that the resulting MS are diagonally misaligned to Pan by half pixel only.

3. Even alignment is totally compensated by means of the even implementation of cubic kernel described here.

4. Even alignment is totally compensated by means of conventional cubic interpolation with odd kernel followed by half-pixel realignment achieved through another odd interpolation by two and selection of the diagonal polyphase component. The integer pixel shift is compensated before or after the second stage of interpolation.

From the values of performance indices in Table 4.5 the following highlights arise:

1. A (diagonal) shift by 1.5 pixels undermines the performances of the AWT fusion method and should be avoided.

2. A half-pixel (diagonal) shift can be tolerated, at least for simulated data that are perfectly superimposed. In simulations with real geocoded

TABLE 4.5: Performance scores of the quality/distortion indices for different interpolation procedures of even-aligned data outlined in the text.

	Case 1	Case 2	Case 3	Case 4
SAM	5.4434	4.7405	4.6769	4.6633
ERGAS	4.5161	3.3559	3.2120	3.1957
Q4	0.8426	0.9029	0.9084	0.9093

products, a certain space-varying misalignment of extent generally lower than one pixel can be noted, mainly because the MS and Pan data are separately projected on cartography each at its own scale. Such a misalignment may be reduced to less than one-half pixel if the cartographic projection is enforced by a DEM of the scene. Therefore, to avoid extra shifts in real cases, also half-pixel shifts stemming from interpolation should be avoided.

3. The proposed bicubic interpolation with an even-sized filter attains the performance of the case, in which odd cubic interpolation is followed by half-pixel bicubic resampling. This procedure requires approximately twice the number of operations, with a customized implementation. Indeed, the performance is slightly better than that attained by the single step even interpolation. An explanation is that the sampling of a *sinc*-like function is more accurate with an odd number of samples, which comprises the maximum, while an even number of samples does not (see Figure 4.6).

A simulation is now presented to better understand the effects of interpolation on pansharpening. Odd-aligned and even-aligned datasets were interpolated each with the correct type of filters, that is, with odd and even filters, respectively. Thus, the interpolated MS image is perfectly overlapped to the 80 cm reference MS available for quality assessment only. SAM, ERGAS, and Q4 were calculated between interpolated MS and reference high-resolution MS and the values are reported in the plots of Figure 4.13. As it appears, between best (odd interpolator of the 11th degree) and worst (even linear interpolator) there is an interval of extent equal to $12 - 15\%$ of the average value, for each index. Such an interval qualifies the importance of interpolation beyond any misregistration issues and explains why NN interpolation, whenever suitable, is preferred to linear interpolation. Compared to the analogous plots in Figure 4.11 relative to the case of the same datasets and same interpolation, in which injection of the spatial details occurs, we note that, unlike ERGAS and Q4, which are substantially improved by fusion, SAM is practically unchanged (slightly lower) after the injection of geometric details, thereby revealing that SAM is inversely related to spectral information and is practically insensitive to the increment in spatial information.

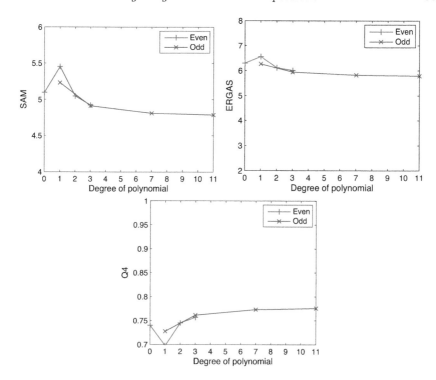

FIGURE 4.13: Quality (Q4) and distortion (SAM, ERGAS) indices of plain interpolation with respect to reference original as a function of the degree of interpolating polynomial. In each plot, images with odd alignment are processed with odd filters and vice versa.

4.4 Concluding Remarks

Although image registration and interpolation should in principle be given comparable emphasis, this chapter mainly focuses on interpolation. In fact, RS data, especially when acquired from the same platform, are preliminarily coregistered by their providers, while the task of interpolation is usually performed by the user. Under this perspective, interpolation as a first step of fusion becomes primary. In fact, preserving the coregistration of the data by means of a correct interpolation contributes to guaranteeing the quality of fusion products, for which geometric correspondence of data from different sensors is essential.

After an overview of registration techniques, mostly implemented in RS software packages, image interpolation is investigated by demonstrating how visual and quantitative performances of fusion methods largely depend on interpolation accuracy and on the presence of possible systematic shifts between expanded MS bands and Pan image.

It is noteworthy that bicubic interpolation, which represents the best trade-off between accuracy and computation, is feasible in a separable fashion, not only with a conventional zero-phase filter of odd length, but also with linear phase filter of even length. This is important because the latter exactly matches the alignment of MS and Pan samples featured by commercial data products and does not require subpixel registration to Pan of interpolated MS, in order to remove the half-pixel shifts originally present between the datasets. In fact, whenever the conventional zero-phase odd interpolator is employed on such datasets, the outcome MS image must be further shifted by half a pixel along both rows and columns. Such an operation, however, requires a further resampling step.

Similar considerations can be made also when the scale ratio r between MS and Pan is not a power of two. Interpolation for the fusion of Hyperion HS data with ALI Pan data $(r = 3)$ is not crucial, because in this case the acquisition geometry of instruments yields each HS pixel perfectly overlapped with the center of a 3×3 block of Pan pixels. Any of the filters reported in the last column of Table 4.4 can be used for interpolation of HS data. Instead, the data from the upcoming PRISMA mission will feature HS at 30 m and Pan at 5 m. A scale ratio $r = 6$ will require one step of odd interpolation by $L = 3$ and one step of even interpolation by $L = 2$, in any order, to match the acquisition geometries of the imaging spectrometer and of the Pan scanner.

Whenever the scale ratio between MS and Pan pixel sizes is a rational number, namely p/q, with p, q integers and $p > q$, interpolation is achieved as the cascade of an interpolation by p followed by a decimation by q [6]. However, MS and Pan may come from different platforms and have a scale ratio that cannot be approximated by the ratio of two integers reasonably small. As an example $7/5 < \sqrt{2} < 10/7$: a better approximation with $\sqrt{2} \approx p/q$ may require p and q, such that interpolation by q is achieved by an extremely long filter. In this case, it is recommended to approach interpolation as a more general problem of resampling, as for all transformations involving a rotation, and not as a linear convolution by a FIR kernel.

Eventually, we recall that polynomial kernels of even size may be implemented also for degrees higher than three. However, border junction conditions of the continuous-time functions must be stated on mid-sample values for kernels of even degrees [88]. This makes the design of higher degree kernels, especially of even degrees, extremely cumbersome, yet less rewarding in terms of added performances.

Chapter 5

Multiresolution Analysis for Image Fusion

5.1 Introduction

Image fusion techniques have been designed to allow integration of different information sources into a unique data product and to also take advantage of the complementary spatial and spectral resolution characteristics typical of most of remote sensing (RS) imaging instruments. This is the case of pansharpening, that is, the spatial enhancement of a multispectral (MS) or a hyperspectral (HS) image through a panchromatic (Pan) observation of the same scene, acquired with a spatial resolution greater than that of the MS/HS image. Since pioneering pansharpening techniques [66] have demonstrated that the spectral distortion of fusion products can be mitigated, if fusion methods based on injecting high-frequency components into resampled versions of the MS data are adopted, many different schemes have been developed by applying different multiresolution approaches.

The goal of this chapter is to demonstrate that multiresolution analysis

(MRA) is a unifying framework in which existing image fusion schemes can be accommodated and novel schemes devised and optimized. The main types of MRA, those suitable for image fusion applications, are described. The basic definitions and principles of MRA are reported in Section 5.2 with specific reference to the dyadic or *octave* case. The focus is on wavelets and their properties in Sections 5.2.1–5.2.2. Wavelet frames provided by overcomplete representations are presented in Sections 5.4.1–5.4.2. When the spatial scale ratio between MS and Pan images is not a power of two, but is any integer or even a fractional number, there are other types of MRA that can provide a versatile solution, as shown in Sections 5.4.2 and 5.5.1.

Unlike separable wavelets that can capture only limited directional structures, nonseparable MRAs contain basis elements *oriented* on a variety of directions. The result is a scale-space analysis with directional sensitivity. This specific capability, described in Section 5.6, has been exploited to develop image fusion methods having high sensitivity to oriented spatial details.

5.2   Multiresolution Analysis

The theoretical fundamentals of MRA will be briefly reviewed in this section with specific reference to the dyadic case, that is, an analysis whose scales vary as powers of two. Thus, the outcome frequency bands exhibit an octave structure, that is, their extent doubles with increasing frequency. This constraint may be relaxed to allow more general analyses [49], as it will be shown in Section 5.5.

Let $L^2(\mathbb{R})$ denote the Hilbert space of real square summable functions, with a scalar product $< f, g > = \int f(x)g(x)dx$. MRA with J levels of a continuous signal f having finite energy is a projection of f onto a basis $\{\phi_{J,k}, \{\psi_{j,k}\}_{j \leq J}\}_{k \in \mathbb{Z}}$ [81].

Basis functions $\phi_{j,k}(x) = \sqrt{2^{-j}}\phi(2^{-j}x - k)$ result from translations and dilations of a same function $\phi(x)$ called the *scaling* function, verifying $\int \phi(x)dx = 1$. The family $\{\phi_{j,k}\}_{k \in \mathbb{Z}}$ spans a subspace $V_j \subset L^2(\mathbb{R})$. The projection of f onto V_j gives an *approximation* $\{a_{j,k} = < f, \phi_{j,k} >\}_{k \in \mathbb{Z}}$ of f at the scale 2^j.

Analogously, basis functions $\psi_{j,k}(x) = \sqrt{2^{-j}}\psi(2^{-j}x - k)$ are the result of dilations and translations of the same function $\psi(x)$ called the *wavelet* function, which fulfills $\int \psi(x)dx = 0$. The family $\{\psi_{j,k}\}_{k \in \mathbb{Z}}$ spans a subspace $W_j \subset L^2(\mathbb{R})$. The projection of f onto W_j yields the wavelet coefficients of f, $\{w_{j,k} = < f, \psi_{j,k} >\}_{k \in \mathbb{Z}}$, representing the *details* between two successive approximations: the data to be added to V_{j+1} to obtain V_j. Hence, W_{j+1} is the complement of V_{j+1} in V_j:

$$V_j = V_{j+1} \oplus W_{j+1}. \tag{5.1}$$

The subspaces V_j realize the MRA. They present the following properties [178]:

$$
\begin{cases}
V_{j+1} \subset V_j, \forall j \in \mathbb{Z} \\
f(x) \in V_{j+1} \Leftrightarrow f(2x) \in V_j \\
f(x) \in V_j \Leftrightarrow f(2^j x - k) \in V_0, \ \forall k \in \mathbb{Z} \\
\bigcup_{-\infty}^{+\infty} V_j \text{ is dense in } L^2(\mathbb{R}) \text{ and } \bigcap_{-\infty}^{+\infty} V_j = 0 \\
\exists \ \phi \in V_0 \text{ such that } \{\sqrt{2^{-j}}\phi(2^{-j}x - k)\}_{k \in \mathbb{Z}} \text{ is a basis of } V_j \\
\exists \ \psi \in W_0 \text{ such that } \{\sqrt{2^{-j}}\psi(2^{-j}x - k)\}_{k \in \mathbb{Z}} \text{ is a basis for } W_j
\end{cases} \tag{5.2}
$$

Eventually, MRA with J levels yields the following decomposition of $L^2(\mathbb{R})$:

$$L^2(\mathbb{R}) = \left(\bigoplus_{j \leq J} W_j \right) \oplus V_J. \tag{5.3}$$

All functions $f \in L^2(\mathbb{R})$ can be decomposed as follows:

$$f(x) = \sum_k a_{J,k} \tilde{\phi}_{J,k}(x) + \sum_{j \leq J} \sum_k w_{j,k} \tilde{\psi}_{j,k}(x). \tag{5.4}$$

The functions $\tilde{\phi}_{J,k}(x)$ and $\{\tilde{\psi}_{j,k}(x)\}_{j \in \mathbb{Z}}$ are generated from translations and dilations of dual functions, $\tilde{\phi}(x)$ and $\tilde{\psi}(x)$, that are to be defined in order to ensure a perfect reconstruction.

The connection between filter banks and wavelets stems from dilation equations allowing us to pass from a finer scale to a coarser one [81]:

$$\phi(x) = \sqrt{2} \sum_i h_i \phi(2x - i)$$
$$\tag{5.5}$$
$$\psi(x) = \sqrt{2} \sum_i g_i \phi(2x - i)$$

with $h_i = \langle \phi, \phi_{-1,i} \rangle$ and $g_i = \langle \psi, \phi_{-1,i} \rangle$.

Normalization of the scaling function implies $\sum_i h_i = \sqrt{2}$. Analogously, $\int \psi(x)dx = 0$ implies $\sum_i g_i = 0$. MRA of a signal f can be performed with a filter bank composed of a lowpass analysis filter $\{h_i\}$ and a highpass analysis filter $\{g_i\}$:

$$a_{j+1,k} = \langle f, \phi_{j+1,k} \rangle = \sum_i h_{i-2k} a_{j,i}$$
$$\tag{5.6}$$
$$w_{j+1,k} = \langle f, \psi_{j+1,k} \rangle = \sum_i g_{i-2k} a_{j,i}.$$

As a result, successive coarser approximations of f at scale 2^j are provided

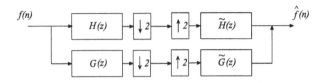

FIGURE 5.1: Dyadic wavelet decomposition (analysis) and reconstruction (synthesis).

by successive lowpass filtering, with a downsampling operation applied to each filter output. Wavelet coefficients at scale 2^j are obtained by highpass filtering an approximation of f at the scale 2^{j-1}, followed by a downsampling.

The signal reconstruction is directly derived from (5.1).

$$
\begin{aligned}
a_{j,k} &= < f, \phi_{j,k} > \\
&= \sum_i \tilde{h}_{k-2i} a_{j+1,i} + \sum_i \tilde{g}_{k-2i} w_{j+1,i}
\end{aligned}
\tag{5.7}
$$

where the coefficients $\{\tilde{h}_i\}$ and $\{\tilde{g}_i\}$ define the synthesis filters.

If the wavelet analysis is applied to a discrete sequence, the original signal samples, $\{f_n = f(nX)\}$, with $X = 1$, are regarded as the coefficients of the projection of a continuous function $f(x)$ onto V_0. The coefficients relative to the lower-resolution subspace and to its orthogonal complement can be obtained through the subsampling of the discrete convolution of f_n by the coefficients of the impulse response of the two digital filters $\{h_i\}$ and $\{g_i\}$, lowpass and highpass, respectively [178]. The two output sequences represent a smoothed version of $\{f_n\}$, or *approximation*, and the rapid changes occurring within the signal, or *detail*.

To achieve reconstruction of the original signal, the coefficients of the approximation and detail signals are upsampled and filtered by the dual filter of $\{h_i\}$ and $\{g_i\}$, or *synthesis* filters $\{\tilde{h}_i\}$ and $\{\tilde{g}_i\}$, which are still lowpass and highpass filters, respectively. The scheme of a wavelet coefficient decomposition and reconstruction is depicted in Figure 5.1, in which $\{f_n\}$ is a discrete 1-D sequence and $\{\hat{f}_n\}$ the sequence reconstructed after the analysis/synthesis stages. As can be seen, the wavelet representation is closely related to a subband decomposition scheme [255].

A sequence $\{\hat{f}_n \equiv f_n\}$ can be reconstructed from the wavelet subbands by using the synthesis filters $\{\tilde{h}_i\}$ and $\{\tilde{g}_i\}$.

5.2.1 Orthogonal Wavelets

The functions $\psi(x)$ and $\phi(x)$ can be constructed in such a way to realize an orthogonal decomposition of the signal; then W_{j+1} is the orthogonal complement of V_{j+1} in V_j. These filters cannot be chosen independently of each other if *perfect reconstruction* (PR) is desired. The synthesis bank must be composed by filters having an impulse response that is a reversed version of that of the analysis ones [178, 255], that is, $\tilde{h}_n = h_{-n}$ and $\tilde{g}_n = g_{-n}$. Quadrature mirror filters (QMF) satisfy all these constraints [80, 81] with $g_n = (-1)^{n-1}h_{-n}$; hence, $G(\omega) = H(\omega + \pi)$. Thus, the *power-complementary* (PC) property, stated in the frequency domain as $|H(\omega)|^2 + |G(\omega)|^2 = 1$, which allows cancellation of aliasing created by downsampling in the dyadic analysis/synthesis scheme shown in Figure 5.1, becomes $|H(\omega)|^2 + |H(\omega + \pi)|^2 = 1$. Despite the mathematical elegance of the decomposition, constraints imposed on QMF do not allow the design of filters with impulse response symmetric around the zeroth coefficient, that is, with null phase, since the number of coefficients is necessarily *even*. Furthermore, the bandwidth value is fixed to be exactly one-half (in the dyadic case) of the available one.

5.2.2 Biorthogonal Wavelets

If the orthogonality constraint is relaxed, we can have symmetric (zero-phase) filters, which are suitable for image processing. Furthermore, the filters of the bank are no longer constrained to have the same size and may be chosen independently of each other. In order to obtain PR, two conditions must be met on the conjugate filters of the filter bank [74]:

$$H(\omega)\tilde{H}^*(\omega) + G(\omega)\tilde{G}^*(\omega) = 1,$$

$$H(\omega + \pi/2)\tilde{H}^*(\omega) + G(\omega + \pi/2)\tilde{G}^*(\omega) = 0.$$

$$(5.8)$$

The former implies a correct data restoration from one scale to another, the latter represents the compensation of recovery effects introduced by downsampling, that is, the *aliasing* compensation. Synthesis filters are derived from the analysis filters with the aid of the following relations:

$$\begin{aligned}
\tilde{h}_n &= (-1)^{n+1}g_{-n} \\
\tilde{g}_n &= (-1)^{n+1}h_{-n}.
\end{aligned}$$

$$(5.9)$$

Biorthogonal wavelets, both in decimated and undecimated versions, are extremely popular for image processing.

5.3 Multilevel Unbalanced Tree Structures

The wavelet decomposition may be recursively applied, that is, the low-pass output of the wavelet transform may be further decomposed into two sequences. This process creates a set of *levels of wavelet decomposition* that represent the signal viewed at different scales. If the decomposition of the low-pass signal is repeated J times, $J + 1$ sequences are obtained: one sequence represents the *approximation* of the original signal containing a fraction $(1/2^J)$ of the original spectrum around zero; the other J sequences are the detail information that allow to reconstruct the original signal. The resulting structure constitutes the unbalanced tree of the discrete wavelet transform (DWT).

A different approach, the so-called wavelet packet decomposition (WPD), can be considered in alternative, by decomposing both the approximation and detail coefficients at each scale to create a full binary tree [179]. However, WPD has been seldom used in remote sensing image fusion, at least successfully.

5.3.1 Critically Decimated Schemes

The decomposition scheme of a signal into three levels ($J = 3$) is shown in Figure 5.3. This representation will be referred to as a *decimated*, or *critically subsampled*, wavelet. In the decimated domain, $a_{j,n}$ and $w_{j,n}$ denote the *approximation*, that is, lowpass, and the *detail*, that is, highpass or bandpass, sequences at the output of the jth stage, respectively.

DWT-based fusion schemes have been proposed in pioneering MRA-based pansharpening methods [165, 272, 106, 278, 217].

5.3.2 Translation Invariant Schemes

The MRA described above does not preserve the translation invariance. In other words, a translation of the original signal samples does not necessarily imply a translation of the corresponding wavelet coefficients.

An equivalent representation is given in Figure 5.2, obtained from that of Figure 5.3 by shifting the downsamplers toward the output and by using upsampled filters [255]. The coefficients before the last downsamplers will be denoted with $\tilde{a}_{j,n}$ and $\tilde{w}_{j,n}$ and this representation will be referred to as an *undecimated*, or *oversampled*, discrete wavelet transform (UDWT).

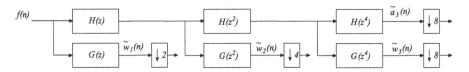

FIGURE 5.2: Three-level scheme ($J = 3$) for undecimated wavelet decomposition (subbands denoted with a tilde).

Note that the coefficients $a_{j,n}$ ($w_{j,n}$) can be obtained by downsampling $\tilde{a}_{j,n}$ ($\tilde{w}_{j,n}$) by a factor 2^j. PR is achieved in both cases. In the undecimated domain, lowpass and highpass coefficients are obtained by filtering the original signal. In fact, from Figure 5.2 it can be noted that at the jth decomposition level, the sequences $\tilde{a}_{j,n}$ and $\tilde{w}_{j,n}$ can be obtained by filtering the original

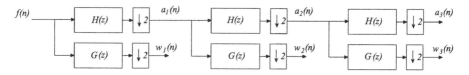

FIGURE 5.3: Three-level scheme $(J = 3)$ for decimated wavelet decomposition.

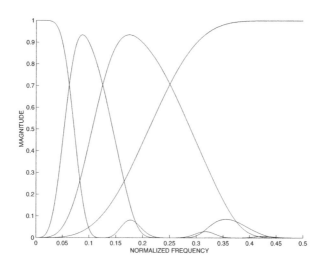

FIGURE 5.4: Frequency responses of the equivalent analysis filters of an undecimated wavelet decomposition, for $J = 3$.

signal f_n through a bank of *equivalent filters* given by

$$H_j^{eq}(\omega) = \prod_{m=0}^{j-1} H(2^m\omega),$$

$$G_j^{eq}(z) = \left[\prod_{m=0}^{j-2} H(2^m\omega)\right] \cdot G(2^{j-1}\omega). \tag{5.10}$$

The frequency responses of the equivalent analysis filters are shown in Figure 5.4. As it appears, apart from the lowpass filter (leftmost), all the other filters are bandpass with bandwidths roughly halved as j increases by one. In Figure 5.4 the prototype filters h and g are Daubechies-4 [80] with $L = 8$ coefficients.

The shift variance of the DWT can be overcome also by applying the Beylkins algorithm [47]: a shift-invariant wavelet transform (SIDWT) can be obtained by computing the wavelet coefficients for all the possible shifts of the signal. This realization considerably reduces the computational complexity of the algorithm. In the 2-D case, at each scale, the DWT is computed on the original image and three circularly shifted versions of the image: (1) all rows shifted by one; (2) all columns shifted by one; (3) all rows and columns shifted by one. This shift-invariant MRA has been applied to pansharpening in Pradhan *et al.* [214].

5.4   2-D Multiresolution Analysis

The wavelet theory for 1-D signals can be easily generalized to the 2-D case. The image signal is a finite energy function $f(x,y) \in L^2(\mathbb{R}^2)$ and a multiresolution approximation of $L^2(\mathbb{R}^2)$ is a sequence of vector subspaces of $L^2(\mathbb{R}^2)$ which satisfies a 2-D extension of the properties (5.2).

In the particular case of separable multiresolution approximations of $L^2(\mathbb{R}^2)$, each vector space V_{2^j} can be decomposed as a tensor product of two identical subspaces of $L^2(\mathbb{R})$, according to

$$V_{2^j} = V_{2^j}^1 \otimes V_{2^j}^1. \tag{5.11}$$

Figure 5.5 illustrates the decomposition scheme to obtain the approximation image A_{j+1} and the three detail images W_{j+1}^{LH}, W_{j+1}^{HL}, and W_{j+1}^{HH} at the scale $j+1$, from the image A_j at the j-th scale. First, the image is decomposed along rows, then along columns. W_{j+1}^{LH}, W_{j+1}^{HL}, and W_{j+1}^{HH} denote the horizontal, vertical, and diagonal details, respectively.

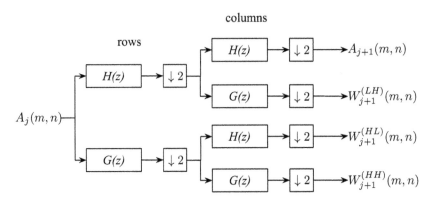

FIGURE 5.5: Decomposition of an image A_j into the approximation image A_{j+1} and the horizontal, vertical, and diagonal detail images W_{j+1}^{LH}, W_{j+1}^{HL}, and W_{j+1}^{HH}, respectively.

The partition of spatial frequency plane achieved by the flowchart in Figure 5.5 is portrayed in Figure 5.6. An example of critically decimated, nonstationary decomposition of an image is reported in Figure 5.7.

This analysis is nonstationary as a direct consequence of the downsam-

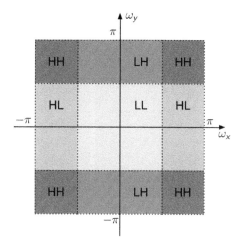

FIGURE 5.6: Spatial frequency plane partition of DWT and UDWT: the four subbands of a depth-one decomposition ($J = 1$) correspond to as many regions labeled as LL, HL, LH, and HH, depending on the filters (L = lowpass, H = highpass) applied along the x and y axes of the image plane, respectively.

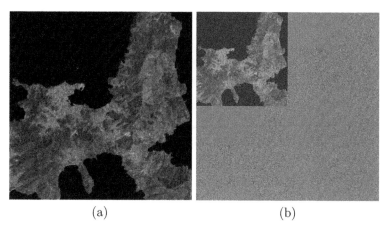

(a)   (b)

FIGURE 5.7: Test RS image (Landsat 4 TM Band #5) of Elba Island, Italy: (a) original; (b) DWT with $J = 1$.

pling operation following each filtering stage. Therefore, wavelet coefficients generated by an image discontinuity may arbitrarily disappear.

5.4.1 2-D Undecimated Separable Analysis

The translation-invariance property is essential in image processing. In order to preserve the translation invariance property, some authors have in-

FIGURE 5.8: UDWT with $J = 1$ of the RS image in Figure 5.7(a). The UDWT coefficients are four times the number of pixels in the original image.

troduced the concept of *stationary* wavelet transform (SWT) [194]. The downsampling operation is suppressed but filters are *upsampled* by 2^j, that is, dilated by inserting $2^j - 1$ zeroes between any couple of consecutive coefficients:

$$h_k^{[j]} = h_k \uparrow 2^j = \begin{cases} h_{k/2^j}, & k = 2^j m, \text{ if } m \in \mathbb{Z} \\ 0, & \text{else} \end{cases}$$

$$g_k^{[j]} = g_k \uparrow 2^j = \begin{cases} g_{k/2^j}, & k = 2^j m, \text{ if } m \in \mathbb{Z} \\ 0, & \text{else}. \end{cases} \tag{5.12}$$

The frequency response of (5.12) will be $H(2^j \omega)$ and $G(2^j \omega)$, respectively [255]. An example of a stationary decomposition of an image is reported in Figure 5.8.

Image multiresolution analysis was introduced by Mallat [178] in the decimated case. However, the 1-D filter bank used for the stationary wavelet decomposition can still be applied in the 2-D case. Image rows and columns are then filtered separately. Filtering relationships to obtain the level $j + 1$ from the jth level are the following, in which (m, n) stands for pixel position:

$$A_{j+1}(m, n) = \sum_k \sum_l h_k^{[j]} h_l^{[j]} A_j(m + k, n + l)$$

$$W_{j+1}^{LH}(m, n) = \sum_k \sum_l g_k^{[j]} h_l^{[j]} A_j(m + k, n + l)$$

$$W_{j+1}^{HL}(m, n) = \sum_k \sum_l h_k^{[j]} g_l^{[j]} A_j(m + k, n + l)$$

$$W_{j+1}^{HH}(m, n) = \sum_k \sum_l g_k^{[j]} g_l^{[j]} A_j(m + k, n + l) \tag{5.13}$$

where A_j is the approximation of the original image at the scale 2^j, giving the low-frequency content in the subband $[0, \pi/2^j]$. Image details are contained in three high-frequency zero-mean 2-D signals W_j^{LH}, W_j^{HL}, and W_j^{HH}, corresponding to horizontal, vertical, and diagonal detail orientations, respectively. Wavelet coefficients of the jth level provide high-frequency information in the subband $[\pi/2^j, \pi/2^{j-1}]$.

For each decomposition level, in the undecimated case images preserve their original size since downsampling operations after each filter have been suppressed. Thus, such a decomposition is highly redundant. In a J-level decomposition, a number of coefficients $3J + 1$ times greater than the number of pixels is generated.

UDWT-based pansharpening was first proposed in Garzelli and Soldati [120] and Aiazzi *et al.* [20, 12], and successively applied in González-Audícana *et al.* [126], Pradhan *et al.* [214], Buntilov and Bretschneider [53], and Garzelli and Nencini [115].

5.4.2   À-Trous Analysis

The "à-trous" wavelet transform (ATW) [93] is a nonorthogonal multiresolution decomposition defined by a filter bank $\{h_i\}$ and $\{g_i = \delta_i - h_i\}$, with the Kronecker operator δ_i denoting an allpass filter. Such filters are not QMF; thus, the filter bank does not allow PR to be achieved if the output is decimated. In the absence of decimation, the lowpass filter is upsampled by 2^j, as in (5.12), before processing the jth level; hence the name "à trous" which means "with holes." In two dimensions, the filter bank becomes $\{h_k h_l\}$ and $\{\delta_k \delta_l - h_k h_l\}$, which means that the 2-D detail signal is given by the pixel difference between two successive approximations, which all have the same scale $2^0 = 1$. The prototype lowpass filter is usually zero-phase symmetric. For a J-level decomposition, the "à-trous" wavelet accommodates a number of coefficients $J + 1$ times greater than the number of pixels. A 3-level ATW decomposition of a sample RS image is reported if Figure 5.10.

An interesting property of the undecimated domain [12] is that at the

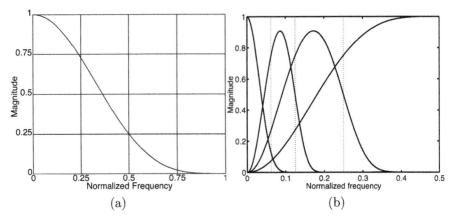

FIGURE 5.9: Frequency responses of lowpass filters h for undecimated MRA: (a) 5-tap Gaussian-like filter introduced by Starck and Murtagh; (b) equivalent filter banks, h_k^* and $g_k^* = \delta(n) - h_k^*$, for $1:4$ analysis, that is, $k = 2$, obtained from the 5-tap prototype by Starck and Murtagh.

kth decomposition level, the sequences of *approximation*, $A_k(n)$, and *detail*, $W_k(n)$, coefficients are straightforwardly obtained by filtering the original signal through a bank of *equivalent filters*, given by the convolution of recursively upsampled versions of the lowpass and highpass filters of the analysis bank:

$$
\begin{aligned}
h_k^* &= \bigotimes_{m=0}^{k-1}(h \uparrow 2^m), \\[2mm]
g_k^* &= \left[\bigotimes_{m=0}^{k-2}(h \uparrow 2^m)\right] \otimes (g \uparrow 2^{k-1}) = h_{k-1}^* \otimes (g \uparrow 2^{k-1})
\end{aligned}
\tag{5.14}
$$

and in the frequency domain:

$$
\begin{aligned}
H_k^*(\omega) &= \prod_{m=0}^{k-1} H(2^m \cdot \omega), \\[2mm]
G_k^*(\omega) &= \left[\prod_{m=0}^{k-2} H(2^m \cdot \omega)\right] \cdot G(2^{k-1} \cdot \omega) = H_{k-1}^*(\omega) \cdot G(2^{k-1} \cdot \omega).
\end{aligned}
\tag{5.15}
$$

ATW has been used extensively in MRA-based pansharpening algorithms, beginning with the method proposed by Núñez *et al.* [196].

5.5 Gaussian and Laplacian Pyramids

The Laplacian pyramid (LP), originally proposed [54] before multiresolution wavelet analysis was introduced, is a bandpass image decomposition

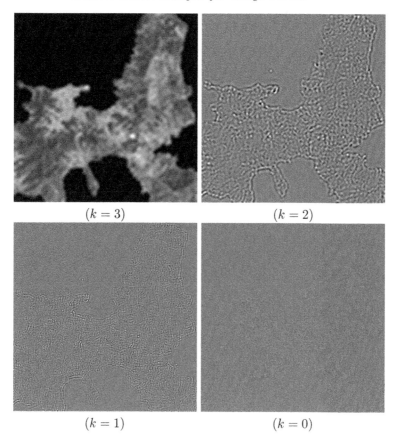

$(k = 3)$ $(k = 2)$

$(k = 1)$ $(k = 0)$

FIGURE 5.10: "À-trous" wavelet transform of a sample RS image; levels 0 through K ($K = 3$) are achieved through the filter bank in Figure 5.9(b).

derived from the Gaussian pyramid (GP), which is a multiresolution image representation obtained through a recursive reduction (lowpass filtering and decimation) of the image dataset.

A modified version of the LP, known as enhanced LP (ELP) [46], can be regarded as an ATW in which the image is recursively lowpass filtered and downsampled to generate a lowpass subband, which is re-expanded and subtracted pixel by pixel from the original image to yield the 2-D detail signal having zero mean. The output of a separable 2-D filter is downsampled along rows and columns to yield the next level of approximation. Again, the detail is given as the difference between the original image and an expanded version of the lowpass approximation. Unlike the baseband approximation, the 2-D detail signal cannot be decimated if perfect reconstruction is desired.

Let $G_0(m,n) \equiv G(m,n)$, $m = 0, \cdots, M - 1$, and $n = 0, \cdots, N - 1$, $M = u \times 2^K, N = v \times 2^K$, be a grayscale image. The classic Burt's GP [54] is defined with a decimation factor of two as

$$G_k(m,n) = \text{reduce}_2[G_{k-1}](m,n)$$

$$\triangleq \sum_{i=-L_r}^{L_r} \sum_{j=-L_r}^{L_r} r_2(i) \times r_2(j) \cdot G_{k-1}(2m+i, 2n+j) \quad (5.16)$$

for $k = 1, \cdots, K$, $m = 0, \cdots, M/2^k - 1$, and $n = 0, \cdots, N/2^k - 1$; k identifies the level of the pyramid, K being the top, or *root*, or *baseband approximation* with size $u \times v$.

The 2D reduction *lowpass* filter is given as the outer product of a linear symmetric kernel, generally odd-sized, that is, $\{r_2(i), i = -L_r, \cdots, L_r\}$ which *should* have the -3 *dB* cutoff at one-half of the bandwidth of the signal, to minimize the effects of *aliasing* [255], although this requirement was not always strictly observed [54].

From the GP, the *enhanced* LP (ELP) [46, 18] is defined, for $k = 0, \cdots, K-1$, as

$$L_k(m,n) \triangleq G_k(m,n) - \text{expand}_2[G_{k+1}](m,n) \quad (5.17)$$

in which $\text{expand}_2[G_{k+1}]$ denotes the $(k+1)$st GP level expanded by two to match the size of the underlying kth level:

$$\text{expand}_2[G_{k+1}](m,n) \triangleq \sum_{\substack{i=-L_e \\ (j+n) \bmod 2=0 \\ (i+m) \bmod 2=0}}^{L_e} \sum_{j=-L_e}^{L_e} e_2(i) \times e_2(j) \cdot G_{k+1}\left(\frac{i+m}{2}, \frac{j+n}{2}\right)$$

$$(5.18)$$

for $m = 0, \cdots, M/2^k - 1$, $n = 0, \cdots, N/2^k - 1$, and $k = 0, \cdots, K-1$.

The 2D lowpass filter for expansion is given as the outer product of a linear symmetric odd-sized kernel $\{e_2(i), i = -L_e, \cdots, L_e\}$, which *must* cut off at one-half of the bandwidth of the signal to reject the *spectral images* introduced by upsampling by two [255]. Summation terms are taken to be null for noninteger values of $(i+m)/2$ and $(j+n)/2$, corresponding to interleaving zeroes. To yield a complete image description, the baseband approximation, that is, $L_K(m,n) \equiv G_K(m,n)$, is taken together with the bandpass ELP.

The attribute enhanced [46] depends on the zero-phase expansion filter being forced to cut off at exactly one-half of the bandwidth [255] and chosen independently of the reduction filter, which may be half-band as well or not. The ELP outperforms Burt's LP and the former LP from Burt and Adelson [54] for many applications, including image compression and fusion, thanks to its layers being almost completely uncorrelated with one another [18].

Figure 5.11 shows the GP and ELP applied on a typical optical remote sensing image. Note the lowpass octave structure of GP layers, as well as the bandpass octave structure of ELP layers. An octave LP is oversampled by a factor of 4/3 at most (when the baseband is one pixel wide). This moderate data overhead is achieved thanks to decimation of the lowpass component. In the case of a scale ratio $p = 2$, that is, frequency octave decomposition, polynomial kernels with three (linear), seven (cubic), 11 (fifth-order), 15 (seventh-order), 19 (ninth-order), and 23 (11th-order) coefficients have been assessed [7]. The term polynomial stems from interpolation and denotes fitting an n-th order polynomial to the nonzero samples. The seven-taps kernel is widespread to yield a bicubic interpolation. It is noteworthy that half-band filters have the even-order coefficients, except the zeroth one, all identically null [255]. The frequency responses of all the filters are plotted in Figure 4.7(a)–(b). Frequency is normalized to the sampling frequency f_S, which is known to be twice the bandwidth available to the discrete signal. The above kernels are defined by the coefficients reported in Tables 4.3 and 4.4 and in Aiazzi *et al.* [12].

The filter design stems from a trade-off between selectivity (sharp frequency cutoff) and computational cost (number of nonzero coefficients). In particular, the absence of ripple, which can be appreciated in the plots with logarithmic scale, is one of the most favorable characteristics.

Pioneering LP-based fusion algorithms have been presented in Aiazzi *et al.* [7, 8, 9, 11] and Alparone *et al.* [35]. Advanced filter bank design for non-dyadic image fusion has been investigated in Blanc *et al.* [49] and Argenti and Alparone [40]. A thorough comparison between DWT and LP fusion is reported in Aiazzi *et al.* [4], varying with the scale ratio of MS to Pan.

5.5.1   Generalized Laplacian Pyramid

When the desired scale ratio is not a power of two, but a rational number, (5.16) and (5.18) need to be generalized to deal with rational factors for reduction and expansion [153].

Reduction by an integer factor p is defined as

$$\text{reduce}_p[G_k](m,n) \triangleq \sum_{i=-L_r}^{L_r} \sum_{j=-L_r}^{L_r} r_p(i) \times r_p(j) \cdot G_k\left(pm+i, pn+j\right). \quad (5.19)$$

The reduction filter $\{r_p(i),\ i = -L_r, \cdots, L_r\}$ must cut off at one pth of bandwidth, to prevent *aliasing* from being introduced. Analogously, an expansion by p is defined as

$$\text{expand}_p[G_k](m,n) \triangleq \sum_{\substack{i=-L_e \\ (j+n)\bmod p=0 \\ (i+m)\bmod p=0}}^{L_e} \sum_{j=-L_e}^{L_e} e_p(i) \times e_p(j) \cdot G_k\left(\frac{i+m}{p}, \frac{j+n}{p}\right) \quad (5.20)$$

The *lowpass* filter for expansion $\{e_p(i),\ i = -L_e, \cdots, L_e\}$ must cut off at one pth of bandwidth. Summation terms are taken to be null for noninteger values of $(i+m)/p$ and $(j+n)/p$, corresponding to interleaved zero samples.

If $p/q > 1$ is the desired scale ratio, (5.16) modifies into the cascade of an

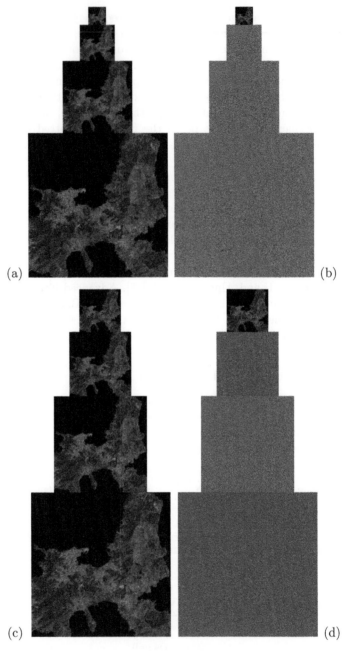

FIGURE 5.11: Examples of (a) GP and (b) ELP, scale ratio $p/q=1/2$; (c) GGP and (d) GLP, scale ratio $p/q=3/2$.

expansion by q and a *reduction* by p, to yield a generalized GP (GGP) [153]:

$$G_{k+1} = \text{reduce}_{p/q}[G_k] \triangleq \text{reduce}_p\{\text{expand}_q[G_k]\} \qquad (5.21)$$

while (5.18) becomes expansion by p followed by reduction by q:

$$\text{expand}_{p/q}[G_k] \triangleq \text{reduce}_q\{\text{expand}_p[G_k]\}. \qquad (5.22)$$

The GLP [7] with p/q scale factor between two adjacent layers, L_k, can thus be defined, for $k = 0, \cdots, K - 1$, as

$$L_k(m, n) \triangleq G_k(m, n) - \text{expand}_{p/q}[G_{k+1}](m, n) \qquad (5.23)$$

which may also be written, by replacing (5.21) in (5.23), as

$$L_k(m, n) \triangleq G_k(m, n) - \text{expand}_{p/q}\{\text{reduce}_{p/q}[G_k]\}(m, n) \qquad (5.24)$$

Again, the baseband approximation is taken into the GLP: $L_K(m, n) \equiv G_K(m, n)$, to yield a complete multiresolution description, suitable for merging data imaged with a p/q scale ratio, p and q being coprime, to yield a unique representation of all possible cases, and thus to avoid designing different and unnecessarily cumbersome filters. Figure 5.11(c)–(d) portrays the GGP and the GLP ($p/q = 3/2$) of the sample image. Example of image fusion with fractional scale ratios are reported in Aiazzi *et al.* [19, 5].

5.6 Nonseparable MRA

As a result of a separable extension from 1-D bases, 2-D wavelet representations are capable of isolating the discontinuities at edge points, but do not efficiently describe the smoothness along the contours. In addition, separable wavelets can capture only limited directional information, which are an important and unique feature of multidimensional signals [87]. Nonseparable MRA not only allows images to be successively approximated, from coarse to fine resolutions using localized coefficients in both the spatial and the scale domains, but it also contains basis elements *oriented* at a variety of directions, much more than the few directions that are offered by separable wavelets. In addition, to capture smooth contours in images, the representation contains basis elements using a variety of elongated shapes with different aspect ratios.

The curvelet transform (CT) [58] is conceptually a multiscale pyramid with many directions and positions at each length scale, and needle-shaped elements at fine scales. The representation provided by countourlet [87] achieves the optimal approximation rate for 2-D piecewise smooth functions with $\mathcal{C}^2$ (twice continuously differentiable) contours.

In image fusion applications, curvelets have been directly applied to pan-sharpening in Garzelli *et al.* [118], Nencini *et al.* [195], and contourlets have been used in a PCA-based scheme in Shah, Younan, and King [229].

An alternative approach to nonseparable MRA is dual-tree complex discrete wavelet transform (DTC-DWT), which has been applied to pansharpening in Ioannidou and Karathanassi [140].

5.6.1 Curvelets

The main benefit of curvelets is their capability of representing a curve as a set of superimposed functions of various lengths and widths. The CT is a multiscale transform but, unlike the wavelet transform, it contains directional elements. Curvelets are based on multiscale ridgelets with bandpass filtering to separate an image into disjoint scales. The side length of the localizing windows is doubled at every other dyadic subband, hence maintaining the fundamental property of the CT, which states that elements of length $2^{j/2}$ are related to the analysis and synthesis of the j-th subband. In practice, the CT consists of applying the block ridgelet transform in Figure 5.12 to the detail frames of the ATW. The algorithm used to implement the digital CT [242] is outlined below:

- Apply the ATW algorithm with J scales. This transform decomposes an image f in its coarse version, c_J, and in the details $\{d_j\}_{\{j=1,...,J\}}$, at scale 2^{j-1}.

- Select the minimum block size, Q_{min}, at the finest scale level d_1.

- For a given scale j, make a partition of d_j in blocks having size

$$Q_j = \begin{cases} 2^{\frac{j}{2}}Q_{min} & \text{if } j \text{ is even} \\ 2^{\frac{j-1}{2}}Q_{min} & \text{if } j \text{ is odd} \end{cases} \tag{5.25}$$

- Apply the ridgelet transform to each block.

Alternatively, the block partition may be replaced with an overlapped set of blocks. The size of blocks is properly increased, for example, doubled for 50% overlap between any two adjacent blocks. During curvelet synthesis, the overlapped regions of detail blocks that are produced are bilinearly interpolated to yield the ATW detail levels of the synthesized image to mitigate possible blocking effects in the reconstructed image.

5.6.2 Contourlets

The nonsubsampled contourlet transform (NSCT) [78], originally introduced as a decimated multiresolution transform in Do and Vetterli [87], is a powerful and versatile tool that allows sparse representation of 2-D signals to

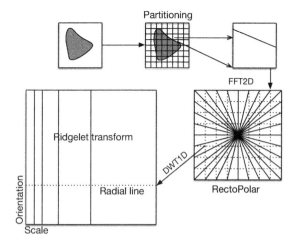

FIGURE 5.12: Flowchart of the discrete ridgelet transform applied to image square blocks.

be achieved. Its main properties are multiresolution, localization, directionality, and anisotropy. In particular, the directionality property permits to resolve intrinsic directional features that characterize the analyzed image, thus overcoming the well-known limitations shown by commonly used separable transforms [211]. Moreover, the NSCT provides also the shift-invariance property, which is essential in the design of image processing algorithms. Another advantage of NSCT over alternative directional transforms is the possibility to define an equivalent filter for subband decomposition.

The NSCT can be seen as a combination of a nonsubsampled pyramid (NSP) and a nonsubsampled directional filter bank (NSDFB). The former yields to the overall system the multiscale, or multiresolution, property, whereas the latter infers a multidirection property. The NSP is implemented by means of a structure similar to Burt's LP [54], with the difference that the NSP does not use any downsampler (upsampler) in the analysis (synthesis) stage. The NSP can also be seen as an equivalent version of the ATW. An example of the frequency splitting performed by the NSP is shown in Figure 5.13(a).

As for the multidirection property, this is achieved by using an NSDFB. Its maximally decimated counterparts have been presented in Bamberger and Smith [43], and Do and Vetterli [87]. The fundamental block of a critically sampled directional filter bank (DFB) is the two-channel quincunx filter bank (QFB). By using suitable combinations of upsampled versions of such filters, wedge shaped supports can be achieved in the frequency domain [87]. An example of an 8-subband splitting is shown in Figure 5.13(b). The important result in the construction of the DFB is that the kth channel can be seen as an equivalent 2-D filter followed by a final equivalent downsampling matrix

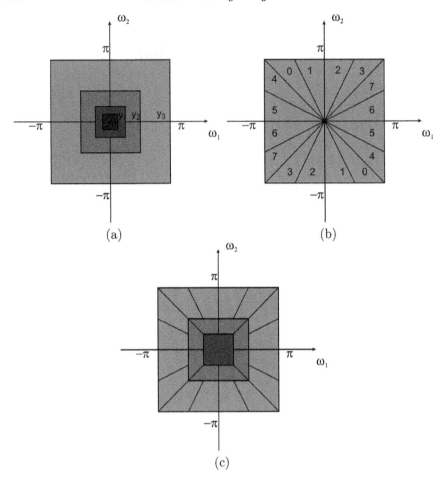

FIGURE 5.13: Spatial frequency analysis of nonsubsampled contourlet transform: (a) octave decomposition of spatial frequency plane; (b) 8-subband directional decomposition of frequency plane; (c) directional octave decomposition achieved by NSCT.

$\mathbf{M}_{k,eq}$. The expression of the final equivalent downsampling matrix can be found in Do [85]. The NSCT is achieved by the cascade of the NSP and the NSDFB. An example of the NSCT frequency splitting is shown in Figure 5.13(c).

Let $A_{J,eq}(\mathbf{z})$ and $B_{J,eq}(\mathbf{z})$ be the lowpass and bandpass equivalent filters that define the NSP, where J is the maximum depth of the NSP and $1 \leq j \leq J$. Let $U^{2^{l_j}}(\mathbf{z})$ be the equivalent filters that implement the NSDFB, where 2^{l_j} is the number of directional output channels realized at the jth multiresolution level. Considering the definitions given above, the equivalent filters of an NSCT can be expressed as

$$H_{J,eq}^{(low)}(\mathbf{z}) = A_{J,eq}(\mathbf{z})$$
$$H_{j,k,eq}(\mathbf{z}) = B_{j,eq}(\mathbf{z}) \, U_{k,eq}^{2^{l_j}}(\mathbf{z})$$

(5.26)

where $1 \leq j \leq J$ and $0 \leq k \leq 2^{l_j}$.

5.7 Concluding Remarks

The main approaches of MRA for image fusion have been presented in this chapter. Several MRA methods have been formally defined, from critically decimated dyadic schemes to overcomplete representations, from simple scale-space decompositions, to more accurate nonseparable scale-space-direction schemes. Such approaches, at least for the separable case, are four:

- DWT: Mallat's DWT is substantially used with success only for data compression. Its drawbacks for fusion are the lack of translation-invariance and the constraint of frequency selective filters to avoid severe aliasing impairments generated by decimated and no longer compensated in the synthesis stage, because coefficients have been altered by fusion between the analysis and synthesis stages [271].

- UDWT: the undecimated version of DWT is stationary and does not incur in aliasing drawbacks. However, the separable processing with mixed lowpass and highpass filters produces planes of wavelet details that do not represent connected sketches of contours at different scales. Such a processing is suitable for decorrelating the data, but is not recommended once one realizes that the best fusion methods are those mimicking the acquisition instrument, whose MTF never mixes lowpass and highpass separable processing.

- ATW: the lack of decimation and of mixed separable filtering allows an image to be decomposed into nearly disjoint bandpass channels in the spatial frequency domain without losing the spatial connectivity of its highpass details, like edges and textures. This property is highly beneficial for fusion. Furthermore, it is easy to design the unique filter of ATW such that integer scale ratios different from powers of two can be tackled.

- GLP: a Laplacian pyramid is substantially a decimated ATW, though not critically. The fact that only the lowpass component is decimated while highpass details are enhanced by fusion makes the behavior of GLP toward aliasing impairments comparable to that of ATW. Furthermore, GLP is suitable for fractional scale ratios between MS and Pan, thanks to the presence of an interpolation stage in its flowchart. Ultimately, GLP-based pansharpening methods are little affected by intrinsic aliasing of the MS image, thanks to the decimation stage of the lowpass component only, which generates opposite aliasing patterns in the details to be injected (see Section 9.3.2).

Nonseparable MRA, like CT and NSCT, can capture directional information at a variety of scales, thus providing image fusion methods with high sensitivity to oriented spatial details. Despite the popularity they encountered a few years ago, the cost / benefits balance is still in favor of ATW / GLP, because nonseparable analysis requires a huge computational effort and, unless a specific nonlinear model of detail-injection is devised and applied in the transformed domain, there is no benefit of using nonseparable MRA [195].

MRA methods are characterized by the prototype lowpass spatial filter that is adopted to design the specific bandpass decomposition. When applied to pansharpening, this prototype filter also defines how to extract the high-spatial frequency information from the Pan image to be injected into the MS image. MRA-based pansharpening algorithms are intrinsically insensitive to spectral distortion, thanks to highpass filtering for spatial detail extraction, but may be unsatisfactory in terms of spatial enhancement. Therefore, spatial filtering in MRA-based pansharpening should be optimized to increase the spatial quality of the pansharpening products, as will be shown in Chapter 7. This spatial optimization should be performed by considering the sensor physical model and the spatial responses of the MS and Pan sensors.

Chapter 6

Spectral Transforms for Multiband Image Fusion

6.1   Introduction

A conspicuous number of image fusion methods proposed in the literature exploits a spectral transformation of the low-resolution multispectral (MS) bands. After interpolating, and possibly coregistering, whenever necessary, the MS data at the scale of the Pan image, a spectral transformation of MS pixels is performed, a *main* component, which must be nonnegative, is replaced with the Pan image, and the inverse spectral transformation yields the sharpened MS bands. Before the substitution, the Pan image is histogram matched to the selected component, which will be denoted as *intensity* component, in order to increase its spectral matching. As will be demonstrated in the following sections in this chapter, this operation is equivalent to inserting the difference between the histogram-matched Pan and the intensity component into the interpolated MS bands to produce the sharpened MS bands [26].

The spectral transformations mostly utilized for sharpening the MS bands are substantially three: the intensity-hue-saturation (IHS) transform, the principal component analysis (PCA) transform, the Gram-Schmidt (GS) transform, derived from the Gram-Schmidt orthogonalization procedure. In the remainder of this chapter, we recall how each of these transformations works and review the most relevant fusion methods based on them.

To better understand the concept of spectral transformation for MS image fusion, we would remind that the aim of the spectral transformation (IHS is the most fitting example concerning spectral bands in the visible wavelengths interval) is to switch from the red-green-blue (RGB) color space, originally introduced to represent and store colors by means of computer systems, to some other space, which is more straightforwardly related to the subjective perception of colors by humans [123].

It is otherwise known that the subjective perception of the intensity component is keener than that of the chromatic components. For this reason, in the digital representation of pictures and video the chromatic components, namely U and V or C_b and C_r are subsampled by four with respect to the luminance component Y without any perceivable degradation. The rationale that a high-resolution black-and-white image, that is, spectrally nonselective in the visible wavelengths, may be combined with two low-resolution chromatic components to produce a high-resolution color image constitutes the foundation of pansharpening fusion. All methods based on spectral transforms developed over the last 25 years were aimed at overcoming the limitation of three bands and the canonical luma-chroma representation of RGB color systems.

6.2   RGB-to-IHS Transform and Its Implementations

Since the first spaceborne MS scanner equipped with a panchromatic instrument, namely SPOT or SPOT-1, the first of the series, has been launched in 1985, the IHS transform has been extensively used for MS image fusion [64, 61, 94], with the limitation that only three bands could be processed at the time. Actually, the blue band was missing in the earlier instruments, like SPOT (up to SPOT 5). The available bands were G, R, and NIR, but the meaning of the transformation is identical.

Analytically, the RGB model is equivalent to a cube in a Cartesian coordinate system [162], where black is on the origin, and red, green, and blue are along the x, y, and z axes, as shown in Figure 6.1, with values normalized between 0 and 1, so that the vertices of the cube lying on these axes indicate pure R, G, and B colors. The diagonal of the cube starting from the origin (black) to the opposite vertex (white) contains pixels having equal R, G, and B values that is perceived as a gray color ranging from black to white. Conversely, the off-diagonal points lying inside the unit cube or on its surface,

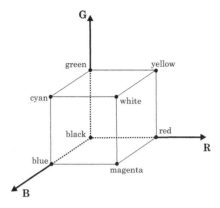

FIGURE 6.1: RGB model in a Cartesian coordinate system with colors in the vertices of the cube.

correspond to chromatic points. In particular, the other three vertices of the cube not belonging to the coordinate axes indicate *yellow*, *cyan*, and *magenta* colors.

IHS was the first transformation used for pansharpening fusion, mainly because the transformed channels relate to some parameters of human color perception, as *intensity* (I), *hue* (H), and *saturation* (S) [144, 172], so that these three components are perceived as *orthogonal* axes in the vision cognitive space. More in details, intensity gives a measure of the brightness of a pixel, that is, the total amount of light, whereas hue refers to the dominant or average wavelength, and for a given pixel is determined by the relative proportions of R, G, and B colors. Saturation specifies the purity of a color in relation to the gray, and depends on the amount of white light mixed with the hue [61]. Consequently, a pure hue is fully saturated. Hue and saturation define the *chromaticity* of a pixel.

6.2.1 Linear IHS Cylindric Transforms

Historically, the IHS model for the color description was first proposed by A. H. Munsell at the beginning of the 20th century, with the aim of applying such a model to the arts and industry [162]. For describing colors, Munsell considered three parameters, because they can mostly reflect the perceptions of a human observer. Originally, he called these parameters *hue*, *value*, and *chroma*, which correspond to hue, intensity, and saturation in the modern versions. The color space devised by Munsell can be described by a cylindrical coordinate system, where the intensity (value) is measured on the axis of the cylinder, the saturation (chroma) is on the axis perpendicular to the z-axis, and the hue is the angle around the z-axis.

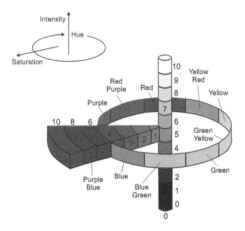

FIGURE 6.2: Munsell's color space in a cylindrical coordinate system described by intensity, hue, and saturation.

To obtain a transformation from the cubic RGB space to the cylindric IHS space, a standard procedure is the following [162]:

- operate a rotation of the RGB color cube described by a Cartesian coordinate system around to the black point (which is already set in the origin of the axes) to obtain an intermediate Cartesian coordinate system (v_1, v_2, I), such that I is the new z axis corresponding to the former diagonal of the cube and represents the pixel intensity level. The former R, G, and B axes are symmetrical arranged around the I axis itself. The v_1 axis is the projection of the former R axis onto the plane orthogonal to the I axis. On this plane, the v_2 axis is orthogonal to the v_1 axis. To fit the previous notation with the current literature, the intermediate Cartesian coordinate system is renamed (I, v_1, v_2).

- operate a conversion of the rotated RGB color cube, i.e., the intermediate Cartesian coordinate system (I, v_1, v_2) to the final IHS cylindrical coordinate system.

There are several versions of cylindrical IHS defined in the literature [28]. Some examples are reported as follows:

- A transformation that is congruent with the previous definition was proposed in Kruse and Raines [154] and implemented in PCI Geomatica. The equations defining the color model are:

$$
\begin{pmatrix} I \\ v_1 \\ v_2 \end{pmatrix} = \begin{pmatrix} \frac{1}{\sqrt{3}} & \frac{1}{\sqrt{3}} & \frac{1}{\sqrt{3}} \\ -\frac{1}{\sqrt{6}} & -\frac{1}{\sqrt{6}} & \frac{2}{\sqrt{6}} \\ -\frac{1}{\sqrt{2}} & \frac{1}{\sqrt{2}} & 0 \end{pmatrix} \begin{pmatrix} R \\ G \\ B \end{pmatrix} \tag{6.1}
$$

$$H = \arctan\left(\frac{v_2}{v_1}\right); \qquad S = \sqrt{v_1^2 + v_2^2} \qquad (6.2)$$

where the superscript $\bullet^T$ denotes matrix transpose, with the inverse transformation defined as:

$$v_1 = S\,\cos(H); \qquad v_2 = S\,\sin(H); \qquad (6.3)$$

$$\begin{pmatrix} R \\ G \\ B \end{pmatrix} = \begin{pmatrix} \frac{1}{\sqrt{3}} & -\frac{1}{\sqrt{6}} & -\frac{1}{\sqrt{2}} \\ \frac{1}{\sqrt{3}} & -\frac{1}{\sqrt{6}} & \frac{1}{\sqrt{2}} \\ \frac{1}{\sqrt{3}} & \frac{2}{\sqrt{6}} & 0 \end{pmatrix} \begin{pmatrix} I \\ v_1 \\ v_2 \end{pmatrix} \qquad (6.4)$$

- A similar transformation, introduced in Harrison and Jupp [132] and reported in Pohl and van Genderen [212] is:

$$\begin{pmatrix} I \\ v_1 \\ v_2 \end{pmatrix} = \begin{pmatrix} \frac{1}{\sqrt{3}} & \frac{1}{\sqrt{3}} & \frac{1}{\sqrt{3}} \\ \frac{1}{\sqrt{6}} & \frac{1}{\sqrt{6}} & -\frac{2}{\sqrt{6}} \\ \frac{1}{\sqrt{2}} & -\frac{1}{\sqrt{2}} & 0 \end{pmatrix} \begin{pmatrix} R \\ G \\ B \end{pmatrix} \qquad (6.5)$$

$$H = \arctan\left(\frac{v_2}{v_1}\right); \qquad S = \sqrt{v_1^2 + v_2^2} \qquad (6.6)$$

and:

$$v_1 = S\,\cos(H); \qquad v_2 = S\,\sin(H); \qquad (6.7)$$

$$\begin{pmatrix} R \\ G \\ B \end{pmatrix} = \begin{pmatrix} \frac{1}{\sqrt{3}} & \frac{1}{\sqrt{6}} & \frac{1}{\sqrt{2}} \\ \frac{1}{\sqrt{3}} & \frac{1}{\sqrt{6}} & -\frac{1}{\sqrt{2}} \\ \frac{1}{\sqrt{3}} & -\frac{2}{\sqrt{6}} & 0 \end{pmatrix} \begin{pmatrix} I \\ v_1 \\ v_2 \end{pmatrix} \qquad (6.8)$$

In the previous cases, the transformation matrices are orthogonal, which means that the inverse matrix is equal to the (conjugate) transpose.

- Other popular cylindrical IHS transformations place the IHS coordinates within the RGB cube, as that reported in Wang *et al.* [264]:

$$\begin{pmatrix} I \\ v_1 \\ v_2 \end{pmatrix} = \begin{pmatrix} \frac{1}{3} & \frac{1}{3} & \frac{1}{3} \\ -\frac{1}{\sqrt{6}} & -\frac{1}{\sqrt{6}} & \frac{2}{\sqrt{6}} \\ -\frac{1}{\sqrt{6}} & \frac{1}{\sqrt{6}} & 0 \end{pmatrix} \begin{pmatrix} R \\ G \\ B \end{pmatrix} \qquad (6.9)$$

$$H = \arctan\left(\frac{v_2}{v_1}\right); \qquad S = \sqrt{v_1^2 + v_2^2} \qquad (6.10)$$

where the inverse transformation is:

$$v_1 = S\,\cos(H); \qquad v_2 = S\,\sin(H); \qquad (6.11)$$

$$\begin{pmatrix} R \\ G \\ B \end{pmatrix} = \begin{pmatrix} 1 & -\frac{1}{\sqrt{6}} & \frac{3}{\sqrt{6}} \\ 1 & -\frac{1}{\sqrt{6}} & -\frac{3}{\sqrt{6}} \\ 1 & \frac{2}{\sqrt{6}} & 0 \end{pmatrix} \begin{pmatrix} I \\ v_1 \\ v_2 \end{pmatrix} \qquad (6.12)$$

- A similar example is reported in Li, Kwok, and Wang [166]:

$$
\begin{pmatrix} I \\ v_1 \\ v_2 \end{pmatrix} = \begin{pmatrix} \frac{1}{3} & \frac{1}{3} & \frac{1}{3} \\ \frac{1}{\sqrt{6}} & \frac{1}{\sqrt{6}} & -\frac{2}{\sqrt{6}} \\ \frac{1}{\sqrt{2}} & -\frac{1}{\sqrt{2}} & 0 \end{pmatrix} \begin{pmatrix} R \\ G \\ B \end{pmatrix}
$$

$$
H = \arctan\left(\frac{v_2}{v_1}\right); \qquad S = \sqrt{v_1^2 + v_2^2} \tag{6.13}
$$

with the inverse transformation given by:

$$
v_1 = S \cos(H); \qquad v_2 = S \sin(H);
$$

$$
\begin{pmatrix} R \\ G \\ B \end{pmatrix} = \begin{pmatrix} 1 & \frac{1}{\sqrt{6}} & \frac{1}{\sqrt{2}} \\ 1 & \frac{1}{\sqrt{6}} & -\frac{1}{\sqrt{2}} \\ 1 & -\frac{2}{\sqrt{6}} & 0 \end{pmatrix} \begin{pmatrix} I \\ v_1 \\ v_2 \end{pmatrix} \tag{6.14}
$$

- Finally, we recall the IHS cylindric transformation reported in Tu *et al.* [252, 251]:

$$
\begin{pmatrix} I \\ v_1 \\ v_2 \end{pmatrix} = \begin{pmatrix} \frac{1}{3} & \frac{1}{3} & \frac{1}{3} \\ -\frac{\sqrt{2}}{6} & -\frac{\sqrt{2}}{6} & -\frac{2\sqrt{2}}{6} \\ \frac{1}{\sqrt{2}} & -\frac{1}{\sqrt{2}} & 0 \end{pmatrix} \begin{pmatrix} R \\ G \\ B \end{pmatrix}
$$

$$
H = \arctan\left(\frac{v_2}{v_1}\right); \qquad S = \sqrt{v_1^2 + v_2^2} \tag{6.15}
$$

with the inverse transformation given by:

$$
v_1 = S \cos(H); \qquad v_2 = S \sin(H);
$$

$$
\begin{pmatrix} R \\ G \\ B \end{pmatrix} = \begin{pmatrix} 1 & -\frac{1}{\sqrt{2}} & \frac{1}{\sqrt{2}} \\ 1 & -\frac{1}{\sqrt{2}} & -\frac{1}{\sqrt{2}} \\ 1 & \sqrt{2} & 0 \end{pmatrix} \begin{pmatrix} I \\ v_1 \\ v_2 \end{pmatrix} \tag{6.16}
$$

All these transformations are linear, that is, they can be described by a transformation matrix. This fact allows us to easily extend the IHS transformation to more than three bands, as in the generalized IHS algorithm (GIHS) [252]. In particular, the transformations described by Eqs. (6.10), (6.13), and (6.15) can be described in a computationally efficient manner for fusion of an arbitrary number of bands.

6.2.2   Nonlinear IHS Transforms

An alternative family of IHS-based transforms is given by the nonlinear IHS, based on the triangle, hexcone, and bi-hexcone models, respectively. In this case, a transformation matrix is not computable to pass from the RGB space to the IHS one.

6.2.2.1 Nonlinear IHS triangle transform

Nonlinear IHS transforms based on the triangle model can be found in Smith [237], Carper, Lillesand, and Kiefer [61], and de Béthune, Muller, and Binard [82]. By crosscutting the RGB cube of Figure 6.1, the IHS space is represented by a triangle model, which is perpendicular to the intensity axis, as in Figure 6.3, with the intensity value also defined as $I = (R + G + B)/3$.

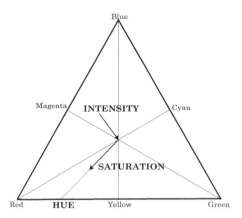

FIGURE 6.3: Nonlinear IHS transform based on the triangle model. The intensity axis is perpendicular to the triangle plane.

Substantially, the triangle itself represents the two-dimensional H-S plane, the H-S coordinates being cyclic, by depending on the positions of the individual colors inside the tree sub-triangles which constitute the RGB triangle [222]. Hue is expressed in a scale ranging from 0 (i.e., blue) to 3 (=0=blue), where the values of 1 and 2 correspond, respectively, to green and red colors. Conversely, saturation values range from 0, that is, no saturation, to 1, that is, full saturation. The nonlinear triangle transformation is given by:

$$I = \frac{R + G + B}{3} \tag{6.17}$$

$$H = \begin{cases} \frac{G-B}{3(I-B)} & \text{if } B = \min(R, G, B); \\ \frac{B-R}{3(I-B)} + 1 & \text{if } R = \min(R, G, B); \\ \frac{R-G}{3(I-B)} + 2 & \text{if } G = \min(R, G, B). \end{cases} \tag{6.18}$$

$$S = \begin{cases} 1 - \frac{B}{I} & \text{if } B = \min(R, G, B); \\ 1 - \frac{R}{I} & \text{if } R = \min(R, G, B); \\ 1 - \frac{G}{I} & \text{if } G = \min(R, G, B). \end{cases} \tag{6.19}$$

6.2.2.2 Nonlinear HSV hexcone transform

The hexcone model tries to transform the RGB colorcube into a set of dimensions modeling an artist's mixing of colors [237], the adoption of a pure hue, or pigment, so that a final tone of that hue can be obtained by adding a mixture of white and black, or a gray color to it. The new dimensions are often called hue, saturation, and value (HSV), and represented by a cone (or a *hexcone*, a figure that can be better named as *hexagonal pyramid*), as represented in Figure 6.4.

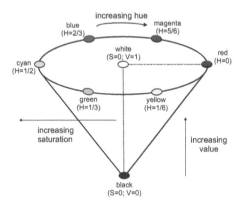

FIGURE 6.4: Nonlinear HSV transform based on the hexcone model. The RGB cube is transformed by modeling an artist's mixing of colors.

In the hexcone model, varying the hue corresponds to traversing the color circle, whereas incrementing the value means to get the image lighter. Finally, if the saturation grows, then we have an increment of the color contribution with respect to that of the gray. Therefore, the H value can be defined both an angle (in degrees) or a value between 0 and 1. In this representation, the values assigned to the RGB primary hues are usually of 0, 1/3, and 2/3, respectively, whereas the secondary hues, yellow, cyan, and magenta take the intermediate values of 1/6, 1/2, and 5/6, respectively. The name hexagonal pyramid just evidences the six vertices that correspond to the primary and secondary colors. Concerning value and saturation, their definitions are:

$$\begin{cases} V = \max(R, G, B); \\ S = \frac{\max(R,G,B) - \min(R,G,B)}{\max(R,G,B)} = \frac{V - \min(R,G,B)}{V}. \end{cases} \qquad (6.20)$$

The hexcone model can be derived as following: if the RGB cube is projected along its main diagonal, that is, the gray axis on a plane perpendicular to the diagonal, a regular hexagonal disk is obtained [237]. For each value of gray, which ranges from 0 (black) to 1 (white), there is a different sub-cube of the RGB cube, to which a projection as before corresponds. By changing the gray levels, each disk is larger than the preceding one, the disk for black being a point, so that the hexcone is formed. Each disk can be selected by varying the V value, so that by specifying V, it means that at least one of R, G, or B equals V, and none is larger, so that $V = \max(R, G, B)$. In particular, if a point has either R, G or B equal to 1, it will have also V equal to 1. Consequently, the hexcone model ignores two components for producing the intensity and will produce equal intensities in the case of a pure color or a white pixel [196].

By considering a disk in the hexcone with the associated V, a single point in the disk is specified by the H and S values. In particular, H represents the angle and S is the length of a vector centered on the gray point of the disk, a length that is relative to the longest possible radius at the given angle, so that S varies from 0 to 1 in each disk. $S = 0$ implies that the color is gray V for the disk determined by V, regardless of hue, whereas $S = 1$ denotes a color lying on the bounding hexagon of disk V, so that at least one of R, G, or B is 0. The derivation of (6.20) for S can be found in Smith [237].

6.2.2.3   Nonlinear HSL bi-hexcone transform

Another nonlinear IHS-based transformation similar to the HSV hexcone model is the HSL bi-hexcone model (hue, saturation, and luminance), where luminance (L) and saturation (S) are defined as follows:

$$\begin{cases} L = \frac{\max(R,G,B)-\max(R,G,B)}{2}; \\ S = \frac{\max(R,G,B)-\min(R,G,B)}{\max(R,G,B)+\min(R,G,B)} & L \le 0.5; \\ S = \frac{\max(R,G,B)-\min(R,G,B)}{2-[\max(R,G,B)+\min(R,G,B)]} & L > 0.5. \end{cases} \qquad (6.21)$$

Similar to the HSV hexcone model, the hue is obtained by traversing the pure color circle and the saturation is described by the distance from the center of the color circle. The difference is that the HSL model can be represented as a double cone (bi-hexcone). Actually, if in the HSV model the number of colors becomes more and more as the value V increases, starting from the black color for $V = 0$, and the maximum is obtained for maximum V (i.e., 1), in the HSL model the maximum number of colors is reached for a lightness of 1/2. Beyond this point, the number of colors again decreases, and for a lightness of 1 a unique color survives, that is, the white color. So, in the HSL case, the intensity produced by a white pixel is 1, whereas the one obtained by a pure color is 1/2.

6.2.3   Generalization of IHS Transforms for MS Image Fusion

This section shows how the IHS transforms described in Sections 6.2.1 and 6.2.2 have been generalized to be used for image fusion methods.

6.2.3.1 Linear IHS

Actually, by disclosing a topic which will be discussed in more detail in the next chapter, an IHS-based fusion algorithm can be characterized by four main steps [264], that is:

- transform the expanded RGB multispectral bands into the IHS components; in this case the intensity image I is obtained by applying the previous relation to the expanded MS band, denoted as $(\tilde{R}, \tilde{G}, \tilde{B})^T$;

- match the histogram of the Pan image with the histogram of the intensity component;

- replace the intensity component with the stretched Pan image;

- apply the inverse IHS transform to obtain the sharpened RGB channels.

Such a procedure is general for each transformation. Consequently, in the following, the Pan image is always intended to be histogram-matched to the image that is substituted before the application of the inverse transform.

As an example, by starting from (6.16), the sharpened MS bands, denoted as $(\hat{R}, \hat{G}, \hat{B})^T$ will be computed as:

$$
\begin{pmatrix} \hat{R} \\ \hat{G} \\ \hat{B} \end{pmatrix} = \begin{pmatrix} 1 & -\frac{1}{\sqrt{2}} & \frac{1}{\sqrt{2}} \\ 1 & -\frac{1}{\sqrt{2}} & -\frac{1}{\sqrt{2}} \\ 1 & \sqrt{2} & 0 \end{pmatrix} \begin{pmatrix} P \\ v_1 \\ v_2 \end{pmatrix}
\tag{6.22}
$$

where the transformed vector $(I, v_1, v_2)^T$ has been obtained by applying (6.15) on the expanded MS bands vector, that is:

$$
\begin{pmatrix} I \\ v_1 \\ v_2 \end{pmatrix} = \begin{pmatrix} \frac{1}{3} & \frac{1}{3} & \frac{1}{3} \\ -\frac{\sqrt{2}}{6} & -\frac{\sqrt{2}}{6} & -\frac{2\sqrt{2}}{6} \\ \frac{1}{\sqrt{2}} & -\frac{1}{\sqrt{2}} & 0 \end{pmatrix} \begin{pmatrix} \tilde{R} \\ \tilde{G} \\ \tilde{B} \end{pmatrix}
\tag{6.23}
$$

By adding and subtracting the I image to the Pan image, we have:

$$
\begin{pmatrix} \hat{R} \\ \hat{G} \\ \hat{B} \end{pmatrix} = \begin{pmatrix} 1 & -\frac{1}{\sqrt{2}} & \frac{1}{\sqrt{2}} \\ 1 & -\frac{1}{\sqrt{2}} & -\frac{1}{\sqrt{2}} \\ 1 & \sqrt{2} & 0 \end{pmatrix} \begin{pmatrix} I + (P - I) \\ v_1 \\ v_2 \end{pmatrix}
\tag{6.24}
$$

If we denote $\delta = P - I$, we finally obtain:

$$
\begin{aligned}
\begin{pmatrix} \hat{R} \\ \hat{G} \\ \hat{B} \end{pmatrix} &= \begin{pmatrix} 1 & -\frac{1}{\sqrt{2}} & \frac{1}{\sqrt{2}} \\ 1 & -\frac{1}{\sqrt{2}} & -\frac{1}{\sqrt{2}} \\ 1 & \sqrt{2} & 0 \end{pmatrix} \begin{pmatrix} I + \delta \\ v_1 \\ v_2 \end{pmatrix} = \\
&= \begin{pmatrix} \tilde{R} \\ \tilde{G} \\ \tilde{B} \end{pmatrix} + \begin{pmatrix} 1 & -\frac{1}{\sqrt{2}} & \frac{1}{\sqrt{2}} \\ 1 & -\frac{1}{\sqrt{2}} & -\frac{1}{\sqrt{2}} \\ 1 & \sqrt{2} & 0 \end{pmatrix} \begin{pmatrix} \delta \\ 0 \\ 0 \end{pmatrix} = \begin{pmatrix} \tilde{R} + \delta \\ \tilde{G} + \delta \\ \tilde{B} + \delta \end{pmatrix}
\end{aligned}
\tag{6.25}
$$

so that $\delta = P - I$ is exactly the high-spatial frequency matrix which is added to the low-resolution MS bands to produce the sharpened products. Consequently, in the case of linear IHS cylindric transforms, the computation of the forward and inverse transformations are actually useless for the fusion procedure, and only the derivation of the intensity I is necessary. This computationally efficient version of the IHS methods is named Fast IHS (FIHS) [252, 269]. In the case of Eqs. (6.10), (6.13), and (6.15), the difference between the Pan and the intensity image is directly injected in the MS bands, whereas for Eqs. (6.2), and (6.6), a multiplicative factor has to be taken into account.

Actually, if the inverse transformation is expressed by means of a generic matrix C, as:

$$\begin{pmatrix} \tilde{R} \\ \tilde{G} \\ \tilde{B} \end{pmatrix} = \begin{pmatrix} c_{11} & c_{12} & c_{13} \\ c_{21} & c_{22} & c_{23} \\ c_{31} & c_{32} & c_{33} \end{pmatrix} \begin{pmatrix} I \\ v_1 \\ v_2 \end{pmatrix} \tag{6.26}$$

then, by substituting the intensity I with the Pan, a more general expression for (6.25) can be obtained:

$$\begin{pmatrix} \hat{R} \\ \hat{G} \\ \hat{B} \end{pmatrix} = \begin{pmatrix} \tilde{R} \\ \tilde{G} \\ \tilde{B} \end{pmatrix} + \begin{pmatrix} c_{11} & c_{12} & c_{13} \\ c_{21} & c_{22} & c_{23} \\ c_{31} & c_{32} & c_{33} \end{pmatrix} \begin{pmatrix} \delta \\ 0 \\ 0 \end{pmatrix} = \begin{pmatrix} \tilde{R} \\ \tilde{G} \\ \tilde{B} \end{pmatrix} + \begin{pmatrix} c_{11} \\ c_{21} \\ c_{31} \end{pmatrix} \cdot \delta \tag{6.27}$$

that is, the first column of the inverse matrix $(c_{11}, c_{21}, c_{31})^T$ represents the vector which multiply the difference $\delta = P - I$ in the fusion procedure. This property may be generalized to any linear transformation.

Note that the substitution of the intensity image with Pan in the fusion procedure allows for the sharpening of the MS bands, but also introduces an annoying color distortion in the fused products, especially if the difference $\delta = P - I$ is large, as in the case of poor overlapping between the spectral responses of the Pan and the three MS bands. This issue is due to the change in the saturation value, whereas the hue is not altered by the IHS fusion [251].

Another advantage of this simplified formulation of the linear IHS fusion procedure is its straightforward extension to more than three bands, which has been proposed in the generalized IHS (GIHS) algorithm [252], in particular by considering the near-infrared band (NIR). For example, a significant reduction of the spectral distortion for images acquired by the IKONOS sensor can be achieved by defining the intensity image as:

$$I = (\tilde{R} + \tilde{G} + \tilde{B} + \tilde{N})/4 \tag{6.28}$$

and then computing the pansharpened bands:

$$\begin{pmatrix} \hat{R} \\ \hat{G} \\ \hat{B} \\ \hat{N} \end{pmatrix} = \begin{pmatrix} \tilde{R} + \delta \\ \tilde{G} + \delta \\ \tilde{B} + \delta \\ \tilde{N} + \delta \end{pmatrix} \tag{6.29}$$

being $\delta = P - I$. Actually, in this case, the spectral response of the Pan image comprises also the spectral response of the NIR band, so that the intensity I of GIHS is more similar to the Pan than the intensity I of IHS.

Obviously, (6.29) can be generalized to K spectral bands, that is, to the vector $(B_1, B_2, \ldots, B_K)^T$, or better to its expanded version $(\tilde{B}_1, \tilde{B}_2, \ldots, \tilde{B}_K)^T$:

$$
\begin{pmatrix} \hat{B}_1 \\ \hat{B}_2 \\ \vdots \\ \hat{B}_K \end{pmatrix} = \begin{pmatrix} \tilde{B}_1 + \delta \\ \tilde{B}_2 + \delta \\ \vdots \\ \tilde{B}_K + \delta \end{pmatrix}
\tag{6.30}
$$

where $\delta = P - I$ and

$$
I = \frac{1}{K} \sum_{i=1}^{K} \tilde{B}_i
\tag{6.31}
$$

Ultimately, we can obtain a more general expression for (6.27), by generalizing it to the case of K bands:

$$
\begin{pmatrix} \hat{B}_1 \\ \hat{B}_2 \\ \vdots \\ \hat{B}_K \end{pmatrix} = \begin{pmatrix} \tilde{B}_1 \\ \tilde{B}_2 \\ \vdots \\ \tilde{B}_K \end{pmatrix} + \begin{pmatrix} c_{11} & c_{12} & \cdots & c_{1K} \\ c_{21} & c_{22} & \cdots & c_{2K} \\ \vdots & \vdots & \vdots & \vdots \\ c_{31} & c_{32} & \cdots & c_{3K} \end{pmatrix} \begin{pmatrix} \delta \\ 0 \\ \vdots \\ 0 \end{pmatrix} =
$$

$$
= \begin{pmatrix} \tilde{B}_1 \\ \tilde{B}_2 \\ \vdots \\ \tilde{B}_K \end{pmatrix} + \begin{pmatrix} c_{11} \\ c_{21} \\ \vdots \\ c_{K1} \end{pmatrix} \cdot \delta
\tag{6.32}
$$

So, for each linear transform, also for K bands, the difference vector $\delta = P - I$ is multiplied for the first column of the matrix C that defines the inverse transform, that is, $(c_{11}, c_{21}, \ldots, c_{K1})^T$ to obtain the pansharpened MS dataset by starting from the expanded one.

Moreover, if the matrix that couples the forward linear transformation is orthogonal, then its inverse is equal to its transpose, that is, the forward transformation is defined by means of the matrix C^T. In this case, the difference vector δ is also multiplied by the first row of the matrix C^T that defines the forward transform.

6.2.3.2 Nonlinear IHS

If the triangle model is dealt with in a framework of a nonlinear IHS fusion algorithm, (6.18) is applied, as usual, to the expanded vector $(\tilde{R}, \tilde{G}, \tilde{B})^T$. Then, by substituting the Pan to I, a backward transformation of IHS to

RGB space with preservation of hue and saturation can be reached, so that the normalized ratio of the new sharpened bands is also preserved, that is:

$$\begin{cases} \frac{\hat{R}}{P} = \frac{\tilde{R}}{I}; \\ \frac{\hat{G}}{P} = \frac{\tilde{G}}{I}; \\ \frac{\hat{B}}{P} = \frac{\tilde{B}}{I}. \end{cases} \qquad (6.33)$$

and, consequently,

$$\begin{cases} \hat{R} = \tilde{R} \cdot \frac{P}{I}; \\ \hat{G} = \tilde{G} \cdot \frac{P}{I}; \\ \hat{B} = \tilde{B} \cdot \frac{P}{I}. \end{cases} \qquad (6.34)$$

By operating some simple manipulations, a new formulation of (6.34), which is comparable with (6.25), can be obtained:

$$\begin{cases} \hat{R} = \tilde{R} + \frac{\tilde{R}}{I} \cdot (P - I) = \tilde{R} + \frac{\tilde{R}}{I} \cdot \delta; \\ \hat{G} = \tilde{G} + \frac{\tilde{G}}{I} \cdot (P - I) = \tilde{G} + \frac{\tilde{G}}{I} \cdot \delta; \\ \hat{B} = \tilde{B} + \frac{\tilde{B}}{I} \cdot (P - I) = \tilde{B} + \frac{\tilde{B}}{I} \cdot \delta. \end{cases} \qquad (6.35)$$

with $\delta = P - I$. Consequently, the equations of the nonlinear IHS triangle model are those of the linear IHS cylindric model weighted by a multiplicative factor which normalizes for each band the details to be injected.

It is noteworthy that the nonlinear IHS triangle model is equivalent to the model underlying the Brovey transform (BT) [122], that is, (6.35) is equal to the equations describing the BT fusion procedure. However, the BT defined by (6.35) and $I = (\tilde{R} + \tilde{G} + \tilde{B})/3$ is not equivalent to the transform defined by (6.18), which is specific for the three R-G-B channels. Actually, BT is independent of the definitions of hue and saturation and can be easily extended to more than three bands, for example, by including the NIR band. It is sufficient to take I as in (6.28), and (6.35) becomes:

$$\begin{cases} \hat{R} = \tilde{R} + \frac{\tilde{R}}{I} \cdot (P - I) = \tilde{R} + \frac{\tilde{R}}{I} \cdot \delta; \\ \hat{G} = \tilde{G} + \frac{\tilde{G}}{I} \cdot (P - I) = \tilde{G} + \frac{\tilde{G}}{I} \cdot \delta; \\ \hat{B} = \tilde{B} + \frac{\tilde{B}}{I} \cdot (P - I) = \tilde{B} + \frac{\tilde{B}}{I} \cdot \delta; \\ \hat{N} = \tilde{N} + \frac{\tilde{N}}{I} \cdot (P - I) = \tilde{N} + \frac{\tilde{N}}{I} \cdot \delta. \end{cases} \qquad (6.36)$$

Then, in the general case of K spectral bands, we have:

$$\begin{cases} \hat{B}_1 = \tilde{B}_1 + \frac{\tilde{B}_1}{I} \cdot \delta; \\ \hat{B}_2 = \tilde{B}_2 + \frac{\tilde{B}_2}{I} \cdot \delta; \\ \vdots \\ \hat{B}_K = \tilde{B}_K + \frac{\tilde{B}_K}{I} \cdot \delta. \end{cases} \qquad (6.37)$$

where $\delta = P - I$ and I is given by (6.31).

By comparing the HSV and HSL models with the linear and triangle models in the framework of image fusion, we can observe that the difference $P - I$ is usually higher in the first case, because the intensity is not a linear combination of the R, G, and B bands, but takes values alternatively from one of the bands, and therefore is less similar to the Pan, which usually overlaps all the visible spectrum. Consequently, it can be expected that the fused images for the HSV and HSL models will be more sharpened but also noisier than for the linear IHS model [73].

6.3 PCA Transform

Principal component analysis (PCA) is a technique extensively used in the image processing field to decorrelate multicomponent datasets, in order to efficiently compact the energy of the input vectors in a reduced number of components of the output vectors.

PCA was first studied by Hotelling for random sequences [136], is the discrete equivalent of the Karhunen-Loeve (KL) series expansion for continuous random processes [141]. The PCA transform is a linear transform which is based on the computation of some data statistics of the second order, as the covariance matrix, of the input data.

6.3.1 Decorrelation Properties of PCA

Consider a dataset constituted by N_p observations, each of K variables, or components, so that usually $N_p \gg K$ holds. The aim of PCA transform is to reduce the dimensionality of the dataset, so that each observation is representable by using a lower number L of variables, that is, $1 \leq L < K$. In the case of the fusion of an MS (HS) dataset, K is the number of bands, usually $K = 4$ for an MS dataset, N_p is the number of pixels of each band, and only the first transformed band is utilized, that is, $L = 1$.

Let the data be arranged as N_p column vectors $X(n) = \{x(m,n), m = 1,\ldots,K\}, n = 1,\ldots,N_p$, each representing the values that the nth pixel takes in the K bands, so that a $K \mathrm{x} N_p$ data matrix X is formed. The matrix X is normalized, that is, the mean values of each row of the original data matrix X_0 (raw data) are computed and stored in a column vector $\mu = \{\mu(m), m = 1,\ldots,K\}$, which is subtracted to each column of the original matrix X_0 to obtain X [240].

The basis vectors of the PCA transform for the zero-mean dataset X are given by the orthonormalized eigenvectors of its covariance matrix C_X. The covariance matrix C_X is defined as:

$$C_X = E\left[XX^T\right] \tag{6.38}$$

being $C_X = C_X(m_1, m_2)$, $m_1 = 1, \ldots, K$, $m_2 = 1, \ldots, K$, that is:

$$C_X(m_1, m_2) = \frac{1}{N_p} \sum_{n=1}^{N} x(m_1, n) \, x(n, m_2) \tag{6.39}$$

Accordingly, C_X is a $K \times K$ matrix whose (m_1, m_2) element is the covariance between the m_1th band and the m_2th band, and, in particular, the mth diagonal element is the variance of the mth band. The Kx1 eigenvectors Φ_i, $i = 1, \ldots K$, where $\Phi_i = \{\phi_{ij}, j = 1, \ldots, K\}$ are given by:

$$C_X \Phi_i = \lambda_i \Phi_i, \quad i = 1, \ldots M \tag{6.40}$$

are columnwise aligned to constitute the PCA transform matrix V, which reduces C_X to its diagonal form. Since C_X is a real symmetric matrix, the matrix V is real orthogonal, because each real symmetric matrix can be orthogonalized by a real orthogonal matrix. Consequently, the inverse of V is equal to its transpose, that is, $V^{-1} = V^T$. By applying the PCA transform matrix, a new $K \times N_p$ data matrix Y is obtained:

$$Y = V^T X \tag{6.41}$$

The inverse transform is given by:

$$X = VY \tag{6.42}$$

where Y has uncorrelated components, that is, its covariance matrix $C_Y = V^T C_X V$ is diagonal [209].

6.3.2   PCA Transform for Image Fusion

In the framework of image fusion, if the PCA is applied to the generic MS dataset $(B_1, B_2, \ldots, B_K)^T$, which is constituted by K bands and expanded to $(\tilde{B}_1, \tilde{B}_2, \ldots, \tilde{B}_K)^T$, each having N_p pixels, a set of more uncorrelated bands $(PC_1, PC_2, \ldots, PC_K)^T$ is computed, by using the following transformation:

$$\begin{pmatrix} PC_1 \\ PC_2 \\ \vdots \\ PC_K \end{pmatrix} = \begin{pmatrix} \phi_{11} & \phi_{12} & \cdots & \phi_{1K} \\ \phi_{21} & \phi_{22} & \cdots & \phi_{2K} \\ \vdots & \vdots & \vdots & \vdots \\ \phi_{K1} & \phi_{K2} & \cdots & \phi_{KK} \end{pmatrix} \begin{pmatrix} \tilde{B}_1 \\ \tilde{B}_2 \\ \vdots \\ \tilde{B}_K \end{pmatrix} \tag{6.43}$$

Since the information is mainly concentrated in the first transformed band, that is, PC_1, by considering (6.32), the set of pansharpened MS band will be given by:

$$\begin{pmatrix} \hat{B}_1 \\ \hat{B}_2 \\ \vdots \\ \hat{B}_K \end{pmatrix} = \begin{pmatrix} \tilde{B}_1 \\ \tilde{B}_2 \\ \vdots \\ \tilde{B}_K \end{pmatrix} + \begin{pmatrix} \phi_{11} \\ \phi_{12} \\ \vdots \\ \phi_{1K} \end{pmatrix} \cdot \delta = \begin{pmatrix} \tilde{B}_1 \\ \tilde{B}_2 \\ \vdots \\ \tilde{B}_K \end{pmatrix} + \begin{pmatrix} \phi_{11} \\ \phi_{12} \\ \vdots \\ \phi_{1K} \end{pmatrix} \cdot (P - PC_1) \tag{6.44}$$

To evaluate the multiplicative term $\Phi_1 = \{\phi_{1i}, i = 1, \ldots, K\}$, let us consider the inverse transform given by (6.42) and take the expression for the generic input $X_i, i = 1, \ldots, K$:

$$X_i = \sum_{j=1}^{K} \phi_{ji} Y_j, \ i = 1, \ldots, K \tag{6.45}$$

By considering the correlation coefficient between the generic input $X_i, i = 1, \ldots, K$ and the generic output $Y_j, j = 1, \ldots, K$, we have:

$$\text{corr}(X_i, Y_j) = \frac{\text{cov}(X_i, Y_j)}{\sqrt{\text{var}(X_i)\,\text{var}(Y_j)}} \tag{6.46}$$

Being X_i and Y_j zero-mean images, their covariance term is given by:

$$\text{cov}(X_i, Y_j) = E[X_i, Y_j] = E\left[\sum_{l=1}^{K} \phi_{li} Y_l, Y_j\right] = \sum_{l=1}^{K} \phi_{li}\, E[Y_l, Y_j] \tag{6.47}$$

By exploiting the fact that Y_l and Y_j are uncorrelated if $l \neq j$, we obtain:

$$\text{cov}(X_i, Y_j) = \phi_{ji}\, E[Y_j, Y_j] = \phi_{ji}\, \text{var}(Y_j) \tag{6.48}$$

and then:

$$\text{corr}(X_i, Y_j) = \frac{\phi_{ji}\, \text{var}(Y_j)}{\sqrt{\text{var}(X_i)\,\text{var}(Y_j)}} = \phi_{ji}\frac{\sqrt{\text{var}(Y_j)}}{\sqrt{\text{var}(X_i)}} = \phi_{ji}\frac{\sigma(Y_j)}{\sigma(X_i)} \tag{6.49}$$

By taking the first transformed band, that is, the first principal component PC_1, and the expanded dataset $(\tilde{B}_1, \tilde{B}_2, \ldots, \tilde{B}_K)^T$, we have:

$$\text{corr}(\tilde{B}_i, PC_1) = \phi_{1i}\frac{\sigma(PC_1)}{\sigma(\tilde{B}_i)} \tag{6.50}$$

Finally, the multiplicative term $\Phi_1 = \{\phi_{1i}, i = 1, \ldots, K\}$ can be evaluated:

$$\phi_{1i} = \text{corr}(\tilde{B}_i, PC_1)\frac{\sigma(\tilde{B}_i)}{\sigma(PC_1)} = \frac{\text{cov}(\tilde{B}_i, PC_1)}{\sigma(\tilde{B}_i)\sigma(PC_1)}\frac{\sigma(\tilde{B}_i)}{\sigma(PC_1)} = \frac{\text{cov}(\tilde{B}_i, PC_1)}{\text{var}(PC_1)} \tag{6.51}$$

Consequently, if the standard MS dataset is considered, that is, $(R, G, B, N)^T$, which has been previously expanded to $(\tilde{R}, \tilde{G}, \tilde{B}, \tilde{N})^T$, (6.44) becomes:

$$\begin{pmatrix} \hat{R} \\ \hat{G} \\ \hat{B} \\ \hat{N} \end{pmatrix} = \begin{pmatrix} \tilde{R} \\ \tilde{G} \\ \tilde{B} \\ \tilde{N} \end{pmatrix} +$$

$$+ \left(\frac{\text{cov}(\tilde{R},PC_1)}{\text{var}(PC_1)}, \frac{\text{cov}(\tilde{G},PC_1)}{\text{var}(PC_1)}, \frac{\text{cov}(\tilde{B},PC_1)}{\text{var}(PC_1)}, \frac{\text{cov}(\tilde{N},PC_1)}{\text{var}(PC_1)} \right)^T (P - PC_1) \tag{6.52}$$

Similar to IHS, it is sufficient to compute the first principal component to achieve the pansharpened bands by the PCA transform.

6.4 Gram-Schmidt Transform

The Gram-Schmidt (GS) transform is based on the Gram-Schmidt orthogonalization procedure [134]. This procedure allows us to build orthogonal bases in Euclidean spaces, both of finite dimensions or not. In the remainder of this section, it will be demonstrated that the application of the GS transform to image fusion includes the PCA transform as a special case.

6.4.1 Gram-Schmidt Orthogonalization Procedure

The Gram-Schmidt orthogonalization procedure can be summarized as follows. Let us consider a set $\{x_1, x_2, \ldots\}$, both finite or not, constituted by vectors in a Euclidean space V. Let $L(x_1, x_2, \ldots, x_k)$ be the subspace generated by the first k elements of this succession. Consequently, a correspondent succession of elements of V exist, that is, $\{y_1, y_2, \ldots, y_k\}$ such that for each k the following properties hold:

- the y_k vector is orthogonal to each element of $L(y_1, y_2, \ldots, y_{k-1})$;

- the subspace generated by $\{y_1, y_2, \ldots, y_k\}$ is the same of the subspace generated by $\{x_1, x_2, \ldots, x_k\}$, i.e., $L(y_1, y_2, \ldots, y_k) = L(x_1, x_2, \ldots, x_k)$;

- the succession $\{y_1, y_2, \ldots, y_k, \ldots\}$ is unambiguously determined except for constant factors possibly different for each k.

Given the set $\{x_1, x_2, \ldots\}$, the orthogonal set $\{y_1, y_2, \ldots\}$ can be obtained by the following recursive formula:

$$y_1 = x_1; \qquad y_{r+1} = x_{r+1} - \sum_{i=1}^{r} \frac{\langle x_{r+1}, y_i \rangle}{\langle y_i, y_i \rangle} y_i, \qquad r = 1, 2, \ldots, k-1 \quad (6.53)$$

where $\langle \bullet, \bullet \rangle$ denotes the scalar product. In particular, if $\{x_1, x_2, \ldots, x_k\}$ is a basis for a Euclidean space of finite dimension, then $\{y_1, y_2, \ldots, y_k\}$ is an orthogonal basis for the same space. Finally, the basis $\{y_1, y_2, \ldots, y_k\}$ can be made orthonormal by dividing each element by its Euclidean norm. Consequently, each Euclidean space of finite dimension can be described by an orthonormal basis.

Note that if x and y are vectors of a Euclidean space, with $y \neq \underline{0}$ (null vector), then the element

$$\frac{\langle x, y \rangle}{\langle y, y \rangle} y$$

is the projection of x along y. This means that the Gram-Schmidt orthogonalization procedure works by subtracting from x_{r+1} all the projections of the same x_{r+1} along the previous orthogonal elements $\{y_1, y_2, \ldots, y_r\}$.

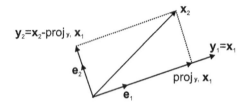

FIGURE 6.5: Example of Gram-Schmidt decomposition in the case of an input dataset composed by two vectors.

The moduli of these projections will be denoted in the following as $\{\text{proj}_{y_1} x_{r+1}, \text{proj}_{y_2} x_{r+1}, \ldots, \text{proj}_{y_r} x_{r+1}\}$. Consequently, we have:

$$\text{proj}_{y_i} x_{r+1} = \frac{\langle x_{r+1}, y_i \rangle}{\langle y_i, y_i \rangle} = \frac{\langle x_{r+1}, y_i \rangle}{\|y_i\|^2}, \qquad i = 1, \ldots, r \qquad (6.54)$$

where $\| \bullet \|$ denotes the Euclidean norm. Figure 6.5 shows an example of Gram-Schmidt decomposition in the simple case of two vectors.

6.4.2 Gram-Schmidt Spectral Sharpening

The basic fusion procedure which exploits the Gram-Schmidt orthogonalization procedure is Gram-Schmidt (GS) spectral sharpening. This technique was invented by Laben and Brower in 1998 and patented by Eastman Kodak [157]. The GS method is widely used since it has been implemented in the Environment for Visualizing Images (ENVI©) software package.

Let I denote the synthetic low-resolution Pan image. By adopting the same notation as for PCA, the input vector for the GS transform, in the case of K bands, will be $X = (I, \tilde{B}_1, \tilde{B}_2, \ldots, \tilde{B}_K)^T$, where the intensity image is obtained by averaging the expanded MS bands:

$$I = \frac{1}{K} \sum_{i=1}^{K} \tilde{B}_i \qquad (6.55)$$

Then, let $Y = (GS_1, GS_2, \ldots, GS_{K+1})^T$ be the output vector after the application of the GS transform. The relationship between X and Y can be expressed in the same manner as (6.41):

$$Y = V^T X \qquad (6.56)$$

with the inverse transform given by:

$$X = VY \qquad (6.57)$$

where V is an orthogonal $(K+1)$ x $(K+1)$ upper triangular matrix. Actually, the columns of V constitute an orthonormal basis, whose nonzero elements, apart from the diagonal ones, come from the GS orthogonalization procedure described in Section 6.4.1, as shown in Lloyd and Bau [175], that is, the generic element (i, j) of V is given by:

$$V(i, j) = v_{ij} = \text{proj}_{y_i} x_j, \ i \geq j \tag{6.58}$$

where

$$\text{proj}_{y_i} x_j = \frac{\langle x_j, y_i \rangle}{\|y_i\|^2}. \tag{6.59}$$

If the vectors are taken all zero-mean, we obtain:

$$\text{proj}_{y_i} x_j = \frac{\text{cov}(x_j, y_i)}{\text{var}(y_i)}. \tag{6.60}$$

By considering the GS orthogonalization procedure and (6.56), it follows that the first transformed band GS_1 coincides with I. Consequently, GS spectral sharpening method substitutes GS_1 with the Pan image, previously histogram-matched to I. On the new output vector Y' the inverse transform is then performed to obtain the pansharpened vector $\hat{X}$, that is:

$$\hat{X} = VY' \tag{6.61}$$

where $\hat{X} = (P, \hat{B}_1, \hat{B}_2, \dots, \hat{B}_K)^T$, that is, the fused bands, apart from the average values, are obtained by starting from the second element of $\hat{X}$.

By taking the (6.32), the set of pansharpened MS band will be given by:

$$\begin{pmatrix} \hat{B}_1 \\ \hat{B}_2 \\ \vdots \\ \hat{B}_K \end{pmatrix} = \begin{pmatrix} \tilde{B}_1 \\ \tilde{B}_2 \\ \vdots \\ \tilde{B}_K \end{pmatrix} + \begin{pmatrix} \text{proj}_{y_1} x_2 \\ \text{proj}_{y_1} x_3 \\ \vdots \\ \text{proj}_{y_1} x_{K+1} \end{pmatrix} \cdot \delta = \begin{pmatrix} \tilde{B}_1 \\ \tilde{B}_2 \\ \vdots \\ \tilde{B}_K \end{pmatrix} + \begin{pmatrix} \text{proj}_I \tilde{B}_1 \\ \text{proj}_I \tilde{B}_2 \\ \vdots \\ \text{proj}_I \tilde{B}_K \end{pmatrix} \cdot \delta \tag{6.62}$$

where $\delta = (P - I)$, the first element of the transformed vector being equal to the first element of the input vector, that is, $y_1 = I$, and $(x_2, x_3, \dots, x_{K+1}) = (\tilde{B}_1, \tilde{B}_2, \dots, \tilde{B}_K)$.

By considering (6.60), we have, for the expanded MS dataset $(\tilde{R}, \tilde{G}, \tilde{B}, \tilde{N})^T$:

$$\begin{pmatrix} \hat{R} \\ \hat{G} \\ \hat{B} \\ \hat{N} \end{pmatrix} = \begin{pmatrix} \tilde{R} \\ \tilde{G} \\ \tilde{B} \\ \tilde{N} \end{pmatrix} + $$
$$+ \left(\frac{\text{cov}(\tilde{R}, I)}{\text{var}(I)}, \frac{\text{cov}(\tilde{G}, I)}{\text{var}(I)}, \frac{\text{cov}(\tilde{B}, I)}{\text{var}(I)}, \frac{\text{cov}(\tilde{N}, I)}{\text{var}(I)} \right)^T \cdot (P - I) \tag{6.63}$$

Consequently, it follows that PCA transform for image fusion is a special case of GS spectral sharpening if we take $I = PC_1$, that is, the intensity image is defined as the first transformed band of the PCA applied to the expanded low-resolution MS dataset.

6.5 Concluding Remarks

In this chapter, the three most important transforms for image pansharpening have been reviewed, starting from the forerunner IHS through PCA to Gram-Schmidt orthogonalization procedure. If a linear transform is addressed, so that its transformation may be stated by means of an invertible square matrix, the fusion procedure can be performed simply by multiplying the difference between the Pan and the first transformed component by the first column of the inverse matrix. In the case of an orthogonal matrix, for example, PCA, the weighting vector is also the first row of the forward matrix.

Consequently, the calculation of the entire transform is useless, because it is sufficient to compute the first transformed band in order to perform the pansharpening procedure. Such simplification allows us to discover remarkable connections among transforms applied to fusion, for example, PCA is a particular case of GS spectral sharpening [26].

Some nonlinear transforms can also be expressed as functions of the first transformed band only, like the nonlinear IHS triangle model, which is equivalent to the Brovey transform. In general, the difference $P - I$ is usually larger for nonlinear than for linear transforms, so that the fused products are usually sharper but may be noisier than those of linear models.

Chapter 7

Pansharpening of Multispectral Images

7.1 Introduction

Pansharpening refers to the fusion of a *panchromatic* (Pan) and a *multispectral* (MS) image acquired on the same area, simultaneously in most cases, but not necessarily. This can be seen as a particular problem of data fusion, since one would aim to combine the spatial details resolved by Pan, but not by MS, and the spectral diversity of MS, which is missing in Pan, in a unique product. With respect to the general problem of multisensor fusion, pansharpening may not require the challenging phase of spatial coregistration, since all images are typically acquired simultaneously, the Pan and MS sensors being both mounted on the same platform. Nowadays, Pan and MS images can be obtained in bundle by several commercial optical satellites such as IKONOS,

OrbView, Landsat 8, SPOT, QuickBird, and WorldView-2. The spatial resolution is even below half meter for the Pan (resulting in the commercial satellite product with the highest spatial resolution) and the spectral resolution
can be up to eight bands captured in the visible and near-infrared (V-NIR)
wavelengths for MS. The fusion of MS and Pan images constitutes the sole
possibility of achieving images with the highest resolutions in both the spatial
and spectral domains. In fact, physical constraints of imaging devices prevent
this goal from being achieved through a single instrument.

The demand of pansharpened data is continuously growing, motivated
by the ever increasing diffusion of commercial products using high-resolution
images, for example, Google Earth and Bing Maps. Furthermore, pansharpening constitutes an important preliminary step aimed at enhancing images
for many RS tasks, such as change detection [241, 50], object recognition
[191], snow mapping [235], thematic classification [52], spectral information
extraction [31], and visual image interpretation [158].

The interest of the scientific community in pansharpening is evident by
reviewing the recent technical literature, in which an increasing number of
contributions on pansharpening can be found. Detailed surveys of pioneering
pansharpening algorithms are available [66, 259, 212]. Critical reviews of the
second generation of pansharpening methods, established with the advent of
a formal multiresolution analysis and objective quality assessments, can also
be found [216, 117, 246, 39, 16, 257]. However, the comparison of the existing
pansharpening methods has been seldom addressed. The contest launched by
the Data Fusion Committee of the IEEE Geoscience and Remote Sensing
Community in 2006 [37] has made a first step in tackling this issue, since
it performed an explicit evaluation of several methods applied on the same
datasets, assessed with the same validation procedure, and using the same
performance metrics.

A further contribution of this chapter is the comparison of the two main
validation procedures used for performance assessment: at full scale and at
a degraded scale. Due to the lack of the reference image (i.e., the unknown
result of the pansharpening), universal measures for evaluating the quality
of the enrichment introduced by pansharpening cannot be explicitly formulated. For this reason, a common practice is the verification of ideal criteria
among which the most credited date back to Wald *et al.* [261]. These criteria
define the characteristics required in the fused product and are formalized
by the *consistency* and *synthesis* properties. The first, which is more easily
achievable in practice, involves the reversibility of the pansharpening process;
it states that the original MS image should be obtainable by simply degrading
the pansharpened image. The synthesis property addresses the characteristics
of the pansharpening result, by requiring that the final product has to reproduce the characteristics of the original MS image at a higher resolution.
This condition entails that both the features of each single channel and the
mutual relations among the bands have to be preserved, justifying the original
articulation in two distinguished statements.

The definition of a technique that fulfills the constraints defined in the protocol is still an open problem [39, 92] and relates closely to the general discussion regarding image quality assessment [262] and image fusion [32], also outside RS [210]. Moreover, there are additional issues linked to the lack of a universally accepted evaluation index matching the human capability in assessing the difference of two images. For example, the mean square error (MSE) has been proven to be inadequate for this task [263], which has given rise to the definition of many other indices for assessing the image quality. In addition, the unavailability of a reference high-resolution MS image precludes the evaluation of the results regardless the chosen quality index. In order to face these aspects and perform a quantitative evaluation of the results, two main solutions have been proposed. The first relies on the reduction of the spatial resolution of both the original MS and Pan images and then the original MS image is used as reference for the evaluation of the results [37]. In this strategy the invariance among scales of the fusion procedures is assumed [261]. However, this hypothesis is not always verified in practice, especially for very high-resolution images acquired on urban areas [32]. The second employs indices that do not require the availability of the reference image [32, 210]. Clearly in this case, the evaluation is done at the scale native of the problem but the results are heavily dependent on the definition of such indices.

7.2 Classification of Pansharpening Methods

Progress in pansharpening methods has been substantially motivated by advances in spaceborne instruments. All instruments launched during the last decade exhibit a ratio of scales between Pan and MS equal to four, instead of two, like in earlier missions, together with the presence of a narrow band in the blue (B) wavelengths and a broadened bandwidth of Pan, also enclosing the near-infrared (NIR) wavelengths. While the change in scale ratios has not substantially influenced the development of fusion methods, the presence of the B band, allowing natural or "true" color display, and of a Pan image that embraces NIR, but not B, to avoid atmospheric scattering, has created significant problems to earlier methods, thereby dramatically motivating the development both of quality assessment tools for pansharpened data, and of alternative fusion methods yielding better quality than earlier ones. In fact, such methods as Intensity-Hue-Saturation (IHS) [61], Brovey Transform (BT) [122], and Principal Component Analysis (PCA) [232] provide superior visual high-resolution MS images but ignore the requirement of high-quality synthesis of spectral information [261]. While these methods are useful for visual interpretation, high-quality synthesis of spectral information is very important for most remote sensing applications based on spectral signatures, such as lithology and soil and vegetation analysis [106].

Over the last two decades, the existing image fusion methods have been classified into several groups. Schowengerdt [226] classified them into spectral domain techniques, spatial domain techniques, and scale space techniques. However, scale space techniques, for example, wavelets, are generally implemented by means of digital filters that are spatial domain techniques. Therefore, methods like HPF [66] and AWL [196], which differ by the type of digital filter, actually belong to the same class.

Ranchin and Wald [217, 246] classified pansharpening methods into three groups: projection and substitution methods, relative spectral contribution methods, and those relevant to the ARSIS concept (Amélioration de la Résolution Spatiale par Injection de Structures), originally employing the (decimated) DWT [217]. It was found that many of the existing image fusion methods, such as HPF [66], GLP [12], and ATW [196], can be accommodated within the ARSIS concept. However, the first two classes, namely "projection and substitution," for example, IHS, and "relative spectral contribution," for example, BT, are equivalent. Section 6.2.3 reviews the work by Tu *et al.* [252], who performed a mathematical development and demonstrated that IHS and PCA do not require the explicit calculation of the complete spectral transformation but only of the component that will be substituted, as happens for BT, which exploits the intensity component only. Therefore, IHS and BT differ only in the way spatial details are weighted before their injection and not in the way they are extracted from the Pan image. Both IHS and BT fusion can be generalized to an arbitrary number of spectral bands.

According to the most recent studies carried out by the authors [45], the majority of image fusion methods can be divided into two main classes. Such classes uniquely differ in the way the spatial details are extracted from the Pan image.

- Techniques that employ linear space-invariant digital filtering of the Pan image to extract the spatial details that will be added to the MS bands [216]; all methods employing multiresolution analysis (MRA) belong to this class.

- Techniques that yield the spatial details as pixel difference between the Pan image and a nonzero-mean component obtained from a spectral transformation of the MS bands, without any spatial filtering of the former. They are equivalent to substitution of such a component with the Pan image followed by reverse transformation to produce the sharpened MS bands [26].

Regardless of how spatial details are obtained, their injection into the interpolated MS bands may be weighed by suitable gains, different for each band, possibly space-varying, that is, a different gain at each pixel. Algorithms featuring context-adaptive, that is, local models generally perform better than schemes based on models fitting each band globally [110, 21, 119, 25]. A pixel-varying injection model is capable of defining fusion algorithms based on spatial modulation of pixel spectral vectors [91], for example, BT for the class of methods based on CS in Figure 7.1 and SFIM [174] for the class of methods based on MRA outlined in Figure 7.4.

The two classes of methods described above exhibit complementary spectral-spatial quality trade-off. Methods without spatial filtering, provide fused images with high geometrical quality of spatial details, but with possible spectral impairments. Methods employing spatial filtering are spectrally accurate in general, but may be unsatisfactory in terms of spatial enhancement. However, if the spectral combination of bands is optimized for spectral quality of pansharpened products [26] and spatial filtering is optimized for spatial quality (MTF filtering yields best results [14]), the two categories yield very similar results in terms of overall quality [25].

These two classes of methods will be detailed in Section 7.3. Approaches different from the two above-mentioned families have also been proposed in the literature. These alternative approaches in some cases are not fully investigated directions or are composed of isolated methods. They are based on regularized solutions of the ill-posed problem consisting in reconstruction of the (unknown) high-resolution image from coarse measurements. Some proposals rely on *Total Variation* penalization terms [203] and others on recent developments in sparse signal representation, or *Compressed Sensing* theory [59, 90]. Among the latter, it is possible to refer to the seminal works [167, 279] that introduced this approach. More recent improvements have been achieved through the application of superresolution techniques, which are already largely diffused in many image processing and computer vision applications [205]. A method belonging to this family has been presented [204]. Also algorithms based on the Bayesian paradigm have been proposed for carrying out the data fusion task [215]. The difficulty in finding a suitable statistical model to jointly characterize the pansharpening result, the available MS and Pan images [98] has strongly limited its use for pansharpening. However, many contributions based on Bayesian estimation theory have been presented in the recent literature. Chapter 11 reviews such new issues in RS image fusion.

Pansharpening has also been proposed for fusing panchromatic and hyperspectral (HS) data [111, 171]. Clearly, this task cannot be addressed by employing conventional methods due to the particular issues that have to be faced (for example, non simultaneous acquisition, coregistration of the data and different spatial coverage and resolution [95, 246]). Chapter 8 is devoted to extending MS pansharpening to HS data.

7.3   A Critical Review of Pansharpening Methods

The goal of this section is to review some of the widely used methods presented in the technical literature over the last years and provide an analysis of their structures and mutual relationships. Due to their wide diffusion and popularity, they can be considered as reference state of the art algorithms for pansharpening. These methods can be grouped into two main classes: CS-based methods and MRA-based methods, plus a further family of *hybrid* methods that employ both CS and MRA. It will be shown that such methods do not constitute a class apart, but fall in the class of MRA-based methods.

The notation and conventions used in the next sections are detailed in the following. Vectors are indicated in bold lowercase: $\mathbf{x}$, with x_i denoting the ith element. Two-dimensional arrays are expressed in bold uppercase: $\mathbf{X} \equiv \{x(i,j)\}_{i=1...,N,j=1,...N}$. Thus, a multiband image is a 3-D array: $\mathbf{X} = \{\mathbf{X}_k\}_{k=1,...,K}$ is a multiband image composed by K bands, in which $\mathbf{X}_k$ indicates the kth band.

7.3.1   Component Substitution

Reportedly, this *spectral* approach to pansharpening relies on the calculation of a spectral transformation of the MS bands, the substitution of a component having all nonnegative values with the histogram-matched Pan image and the calculation of the inverse transformation to yield back the pansharpened MS bands. This family comprises classical algorithms such as *intensity-hue-saturation* (IHS) [61, 66], *principal component analysis* (PCA) [65, 232], and *Gram-Schmidt spectral sharpening* (GS) [157, 26].

This class of methods is based on the projection of the MS image into another vector space, by assuming that the transformation is able to separate the spatial structure and the spectral information of the scene into different components; hence its alternative name of *Projection Substitution* [246]. Subsequently, the component containing the spatial information is substituted by the Pan image. In principle, the greater the correlation between the Pan image and the replaced component, the less spectral distortion will be introduced by this approach [246]. To this purpose, histogram matching of Pan to the selected component is performed before the substitution. The process is completed by bringing the data back to the original MS space through the inverse transformation.

This approach is global (that is, it operates in the same way on the whole image) leading to advantages and limitations. In greater detail, techniques belonging to this family are usually characterized by a high fidelity in rendering the spatial details in the final product [26]; they are generally fast and easy to implement. On the other side, they may be unable to account for local dissimilarities between Pan and MS images, which may produce significant spectral distortions [246, 39].

A new formalization of the CS approach was proposed by Tu *et al.* [252] and then analyzed other subsequent works [251, 91, 26, 72]. It was shown that under the hypothesis of a linear transformation and the substitution of only a single nonnegative component, the fusion process can be obtained without the explicit application of the forward and backward transformations but by a

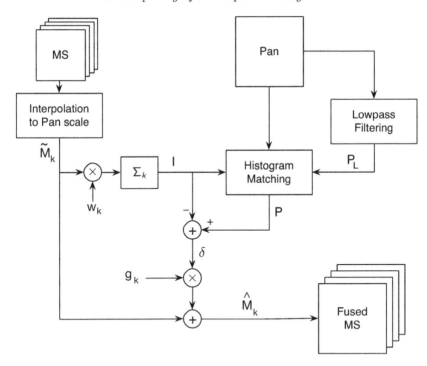

FIGURE 7.1: Flowchart detailing the steps of a generic pansharpening process based on the CS approach. By following the guidelines of Chapter 6 (see (6.32)), the blocks computing the direct and inverse transforms are omitted and the former reduces to a linear combination of bands with weights w_ks, the latter is encapsulated in the injection gains g_ks.

proper injection scheme. This observation allows for a faster implementation of these methods. Starting from this consideration, a general formulation of the CS fusion scheme is given by

$$\hat{\mathbf{M}}_k = \tilde{\mathbf{M}}_k + g_k \left(\mathbf{P} - \mathbf{I}\right), \tag{7.1}$$

in which the subscript k indicates the kth band, $\mathbf{g} = [g_1, \ldots, g_k, \ldots, g_K]$ is the vector of the *injection gains*, while $\mathbf{I}$ is defined as

$$\mathbf{I} = \sum_{i=1}^{K} w_i \tilde{\mathbf{M}}_i, \tag{7.2}$$

where the weights $\mathbf{w} = [w_1, \ldots, w_i, \ldots, w_K]$ may be chosen, for example, to measure the relative spectral overlap between each MS band the Pan image [251, 246].

By following the guidelines of Tu *et al.*, Dou *et al.*, Aiazzi *et al.*, and Choi *et al.* [252, 251, 91, 26, 72] it has been shown that the fusion process in (7.1) can be carried out without the explicit computation of the spectral transformation. This result is reviewed in Sections 6.2.3, 6.3.2, and 6.4.2, for generalized IHS (GIHS), PCA, and GS fusion. Accordingly, the flowchart of the generic CS fusion method is shown in Figure 7.1 and does not include the direct and inverse spectral transformations. Lowpass filtering of Pan is required because the histogram-matched version of Pan, namely, $\mathbf{P}$, after degrading its resolution to the resolution of $\mathbf{I}$, that is, of the original MS bands, should exhibit same mean and variance as $\mathbf{I}$, while in general $\sigma(\text{Pan}) > \sigma(P_L)$, that is, integration over a larger PSF reduces the variance of the imaged radiance.

The steps of the generic CS-based pansharpening are the following:

1. interpolate MS to match the scale of Pan and exactly overlap to it;

2. input the spectral weights $\{w_k\}_{k=1,\dots,K}$;

3. calculate the intensity component according to (7.2);

4. match the histogram of Pan to I: $\mathbf{P} = [\text{Pan} - \mu(\text{Pan})] \cdot \frac{\sigma(I)}{\sigma(P_L)} + \mu(I)$;

5. calculate the band dependent injection gains $\{g_k\}_{k=1,\dots,K}$;

6. inject, i.e., add, the extracted details according to (7.1).

TABLE 7.1: Values of spectral weights in (7.2) and injection gains (7.1) for some CS-based methods (see Section 6.2.3.2 for the g_k of BT).

Method	w_k	$\mathbf{G}_k$
GIHS	$1/K$	1
BT	$1/K$	$\tilde{\mathbf{M}}_k/\mathbf{I}$
PCA	$\phi_{1,k}$	$\phi_{1,k}$
GS	$1/K$	$\frac{\text{cov}(\mathbf{I},\mathbf{M}_k)}{\text{var}(\mathbf{I})}$

It is noteworthy that without the contribution of Tu *et al.* and Dou *et al.* [252, 251, 91], IHS fusion would not be feasible for more than three bands. In Section 7.3.2 it will be shown that an adaptive estimation of the component that shall be substituted with Pan is the key to improve the performances of CS methods (*adaptive CS* [26]).

In the following, a more detailed description of the main CS techniques is presented. The values of the coefficients for these techniques, as indicated in (7.1) and (7.2), are summarized in Table 7.1. In PCA entry, the coefficients $\{\phi_{1,k}\}_{k=1,\dots,K}$ represent the first row of the forward transform, which is also the first column of the inverse transform. The orthonormalization constraint yields $\sum_k \phi_{1,k}^2 = 1$.

7.3.1.1 Generalized Intensity-Hue-Saturation

The term *generalized intensity-hue-saturation* [252] refers to the extension of the original IHS method [61] to an arbitrary number of bands K, with $K > 3$. The procedure is described in Section 6.2.3. Here we wish to state this type of fusion in its most general form, that is, with the most general choice of (nonnegative) spectral weights.

In fact, it may be that MS and Pan data have different dynamics, for example, because they are expressed as *radiance* units, instead of *spectral radiance*, that is, spectral radiance density units. In the former case, Pan is the sum of the MS bands, not their average. Spectral weights that do not sum to one may also occur if the datasets are not radiometrically calibrated. All pansharpening methods contain one step in which the Pan image is cross-calibrated over the component to be replaced, which is a function of the MS bands, as it happens with CS methods, or directly over each of the MS bands, as it happens with MRA methods. Therefore, uncalibrated MS and Pan data may be merged together and the resulting sharpened MS image may be radiometrically calibrated at a subsequent time. Eventually, when pansharpening is extended to datasets having intrinsically different word lengths, like thermal and visible observations, the sum of spectral weights may be far different from one.

Whenever the spectral weights $\{w_k\}_{k=1,\dots,K}$ are constrained neither to be all equal to $1/K$ nor to sum to one, by exploiting the structure of the forward and inverse IHS transform matrices for $K = 3$ (6.27), it is found that the first column of the inverse matrix has all elements equal to $(\sum_l w_l)^{-1}$. Hence, in (7.1) $g_k = (\sum_l w_l)^{-1}$, $\forall k$. Accordingly, the GIHS fusion method may be reformulated as:

$$\hat{\mathbf{M}}_k = \tilde{\mathbf{M}}_k + \left(\sum_{l=1}^{K} w_l \right)^{-1} \cdot (\mathbf{P} - \mathbf{I}), \ k = 1, \dots, K \tag{7.3}$$

in which $\mathbf{P}$ is the histogram-matched Pan and $\mathbf{I}$ follows from (7.2). This approach is also called *fast* IHS (FIHS) [251] because it avoids the sequential computation of the direct transformation, substitution, and the final backward step.

In Section 6.2.3.2 it has been proven that, if nonlinear IHS is used instead of linear IHS, the BT pansharpening method is obtained. Accordingly, the fused image is defined, for $k = 1, \dots, K$, as:

$$\hat{\mathbf{M}}_k = \tilde{\mathbf{M}}_k \cdot \frac{\mathbf{P}}{\mathbf{I}}. \tag{7.4}$$

Since (7.4) can be recast as

$$\hat{\mathbf{M}}_k = \tilde{\mathbf{M}}_k + \frac{\tilde{\mathbf{M}}_k}{\mathbf{I}} \cdot (\mathbf{P} - \mathbf{I}), \tag{7.5}$$

in which again **P** is Pan histogram-matched to **I**. The model (7.1) may be achieved by setting the *space-varying* injection gains given by

$$\mathbf{G}_k = \frac{\tilde{\mathbf{M}}_k}{\mathbf{I}}, \ k = 1, \ldots, K. \tag{7.6}$$

This specific choice of the parameter, together with the values of the weight vector **w**, is reported in Table 7.1.

7.3.1.2   Principal Component Analysis

Section 6.3 described *principal component analysis* (PCA) in detail, another spectral transformation widely employed for pansharpening applications [66, 65, 232]. PCA is achieved through a multidimensional rotation of the original coordinate system of the K-dimensional vector space around the barycenter of the set of spectral pixel vectors, such that the projection of the original vectors onto the new axes, which are the eigenvectors of the covariance matrix along the spectral direction, produces a set of scalar images, called *principal components* (PCs), that are statistically uncorrelated to each other. PCs are generally sorted for decreasing variance, which quantifies their information content.

More specifically, the hypothesis underlying its application to pansharpening is that the spatial information (shared by all the spectral channels) is concentrated in the first component, while the spectral information (peculiar to each single band) is accounted by the other components. Actually, the spatial information is mapped onto the first component, to an extent proportional to the correlation among the MS channels [39]. Again, the whole fused process can be described by the general formulation stated by (7.1), where the **w** and **g** coefficient vectors are derived by the PCA procedure on the MS image; consequently, no specific expression is provided in Table 7.1 for **w** and **g**, since they are dependent on the particular processed dataset.

The main drawback of PCA for pansharpening is that it is never implicit that the first component, PC_1, spectrally matches the Pan image, neither it is possible to define a PCA transform, whose first component is optimized for spectral matching with Pan, as it happens with GIHS and GS (see Section 7.3.2).

7.3.1.3   Gram-Schmidt orthogonalization

The *Gram-Schmidt* (GS) orthogonalization procedure is a common technique used in linear algebra and multivariate statistics to *orthogonalize* a set of nonorthogonal vectors. The GS procedure has been used to achieve a powerful pansharpening method, which was patented by Laben and Brower for Kodak in 2000 [157] and implemented in the ENVI© software package.

In the fusion process, the global mean of each band, that is, the spatial average, is subtracted from each pixel, before the orthogonalization procedure is started. Besides the K spectral bands, interpolated at the scale of Pan and lexicographically ordered and mapped onto vectors having as many components as the number of pixels, a synthetic low-resolution panchromatic image **I** is used, as the first basis vector. The outcome of the orthogonalization procedure is the start-up vector **I**, the component of the first spectral bands that is orthogonal to **I**, the component of the second spectral band that is orthogonal to the plane defined by **I** and by the the first orthogonal component, and so on. Before the inverse transformation is accomplished, **I** is replaced by the histogram-matched Pan, **P**, and the original MS bands are recovered from the new first component and from the orthogonal components. Section 6.4 reviews the main features of GS spectral sharpening.

It is noteworthy that GS constitutes a transformation more general than PCA, which can be obtained from GS if PC_1 is taken as the first component instead of **I** [25]. This suggests building an orthogonal representation of the MS bands starting from their linear combination optimized for spectral matching with Pan.

Again, this process is expressed by (7.1), by using, for $k = 1, \ldots, K$, the gains [26]:

$$g_k = \frac{\text{cov}(\tilde{\mathbf{M}}_k, \mathbf{I})}{\text{var}(\mathbf{I})}, \tag{7.7}$$

in which $\text{cov}(\mathbf{X}, \mathbf{Y})$ indicates the covariance between two images **X** and **Y**, and $\text{var}(\mathbf{X})$ is the variance of **X** (see also Table 7.1).

The simplest way to obtain this low-resolution approximation of Pan consists of simply averaging the MS components of each pixel (that is, in setting $w_i = 1/K$, for all $i = 1, \ldots, K$); this default modality is called *GS Mode 1*. According to the alternative *GS Mode 2*, the low-resolution approximation of Pan **I** is user defined. This entails the use of a different set of spectral weights or the application of a proper lowpass filter to the original Pan image, thereby leading to a hybrid method that exhibits the prominent features of the MRA pansharpening class.

7.3.2 Optimization of CS Fusion Methods

CS fusion methods, either in their classical form, with forward and inverse transformations, or in their alternative form, without explicit calculation of neither transformation, as shown in Figure 7.1, rely on the spectral matching between the synthesized intensity **I** and the lowpass approximation of Pan, $\mathbf{P}_L$, in the assumption that the spectral responsivity of the MS channels is overlapped with that of the Pan channel [124]. The simplest optimization is calculating spectral weights by measuring the relative spectral contributions of the MS bands to the Pan image. This strategy has been accomplished [251] for IKONOS imagery, whose spectral responsivity is reported in Figure 2.12. For the four bands of IKONOS, the optimal spectral weights are $w_1 = 1/12$, $w_2 = 1/4$ and $w_3 = w_4 = 1/3$, roughly corresponding to the relative spectral contributions of MS to Pan, and are obtained as averages of manually adjusted sets of weights over a large number of images of the same instrument on different landscapes.

To understand the influence of the spectral response on the fusion of MS images, let us consider the relative spectral responses of IKONOS, as plotted in Figure 2.12. Ideally, the MS bands (B, G, R, and NIR) should be disjoint and should entirely fall within the bandwidth of Pan. From Figure 2.12, however, it appears that the G and B bands are largely overlapped and that the B band mostly falls outside the 3 dB cutoff of the Pan band. Furthermore, the spectral response of Pan is extended beyond the NIR band. The color distortion problem in fusion stems from such mismatches, and in particular from the fact that an approximation of Pan synthesized as a plain average of the B, G, R, and NIR bands does not match the spectral response in Figure 2.12. As an example, vegetation appears relatively *bright* in the NIR and Pan bands, while its reflectance is low in the visible bands. When all bands are equally weighted, the radiometric value of vegetated areas is likely to be smaller in the synthetic Pan than in the true Pan. This effect causes the injection of a radiance offset into some of the fused MS bands, which may give rise to color distortion, mostly noticeable in true-color display of vegetated regions.

To avoid this inconvenience, the trivial solution is to generate the synthetic approximation of Pan in such a way that the spectral response of the sensor is considered, by differently weighing the contributions coming from the different spectral channels [251, 200, 124]. As a matter of fact, the spectral weights turn out to be landscape-dependent and their optimization should be made image by image, or even class by class on the same image. Hence, a robust solution based on multivariate regression has been proposed [26].

The preprocessing steps to be performed are the following:

1. Spatially degrade the original full-resolution Pan image to the size of the MS bands by means of proper lowpass filtering and decimation.

2. Assume that at every pixel position it holds that:

$$\mathbf{P}_L \downarrow r = \sum_{k=1}^{K} w_k \cdot \mathbf{M}_k + n \qquad (7.8)$$

   in which $\mathbf{P}_L$ is the lowpass-filtered Pan, $\{w_k\}_{k=1,...,K}$ are the (unknown) spectral weights and n is a zero-mean noise.

3. Find the optimal set of weights $\{\hat{w}_k\}_{k=1,...,K}$ such that the mean square value of n is minimized.

4. According to (7.2), use $\{\hat{w}_k\}_{k=1,...,K}$ to synthesize a generalized $\mathbf{I}$ component that can be used in either GS or GIHS.

The implicit assumption is that the regression coefficients computed at the scale of the MS image are equal to those that would be computed at the

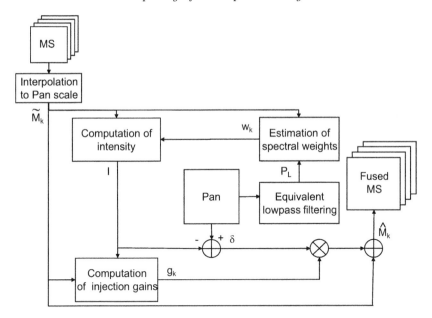

FIGURE 7.2: Optimized CS pansharpening scheme.

scale of the original Pan image, if multiband observations were available. This assumption is reasonable for the *spectral* characteristics of the dataset, but may no longer hold for the *spatial* ones.

The flowchart of the optimized CS fusion, either GS or GIHS, is shown in Figure 7.2. It is noteworthy that the histogram matching of Pan is missing, because the least squares (LS) solution of (7.8) yields optimized weights such that I is implicitly histogram-matched to Pan. Analogously, BT can also be optimized by means of optimal regression-based weights [29].

An optimization of spectral weights based on interband correlation constitutes the *generalized intensity* [36] used for optical and SAR fusion and reviewed in Section 10.3.4.1.

Ultimately, among CS-like optimized methods we can find the *Band-Dependent Spatial-Detail* (BDSD) algorithm [119], which uses an extension of the generic formulation reported in (7.1) and points at estimating separately for each band both **w** and **g** according the MMSE criterion. Due to the fact that this method is not strictly a CS method and is robust to a partial lack of overlap between the spectral channels of MS and Pan, it will be reviewed in Section 8.3.1 concerning pansharpening of HS data.

7.3.3 Multiresolution Analysis

MRA, which represents the *spatial* approach to pansharpening, relies on the injection of spatial details, obtained through a multiresolution decom-

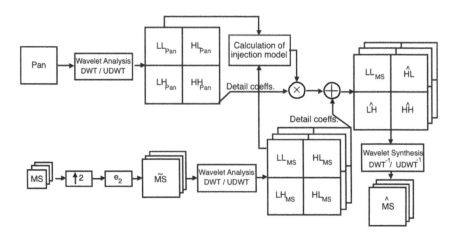

FIGURE 7.3: Flowchart of MRA-based pansharpening exploiting DWT/UDWT. For simplicity of flow, the scale ratio is $r = 2$ and UDWT is employed.

position of the Pan image, into the resampled MS bands [38]. The spatial details can be extracted according to several modalities of MRA: (decimated) *Discrete Wavelet Transform* (DWT) [178], *Undecimated Wavelet Transform* (UDWT) [194], *À-Trous Wavelet Transform* (ATW) [93, 231], *Laplacian Pyramid* (LP) [54], nonseparable transforms, either based on wavelets (for example, contourlet [87]) or not (curvelet [243]). Chapter 5 synthetically reviews the formal definition of MRA and the four types mentioned in this section, with emphasis on image fusion.

Although forerunners of this approach were not based on a formal MRA, like HPF [66] and HPM [226], pioneering MRA-based methods [165, 272, 106, 278, 217] strictly used DWT. Later, the favorable features of UDWT (translation invariance, looser choice of filter banks, absence of aliasing artifacts) led to a brief period of popularity. However, DWT and UDWT were quickly abandoned after ATW and (G)LP were introduced. A notable exception is the recent *indusion* method [151], which profits from multiple equalization steps and achieves interesting results.

In the assumption of a dyadic MRA, the steps of the generic pansharpening procedure for MS and Pan data, whose scale ratio is r, are the following:

1. interpolate MS to match the scale of Pan and exactly overlap to it;

2. calculate the MRA of each of the MS bands with a depth equal to $\log_2(r)$;

3. calculate the MRA of the Pan image with a depth equal to $\log_2(r)$;

4. calculate the band dependent injection gains $\{g_k\}_{k=1,...,K}$, possibly one g_k for each orientation (LH-horizontal, HL-vertical, HH-diagonal);

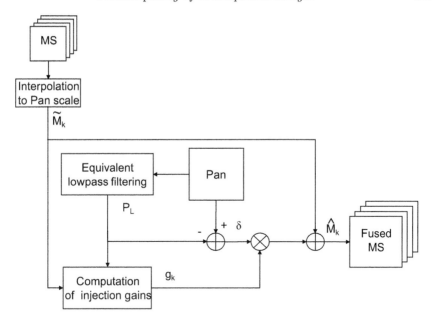

FIGURE 7.4: Flowchart of a generic pansharpening algorithm belonging to the MRA class.

5. add the detail subbands of Pan weighted by the corresponding injection gain to the corresponding subbands of each MS band;

6. apply the inverse transform to the enhanced subbands of each MS band.

Figure 7.3 illustrates the fusion procedure for the case of UDWT and $r = 2$. In Ranchin and Wald [217], this paradigm has been denoted as *Amélioration de la Résolution Spatiale par Injection de Structures* (ARSIS) to highlight that the purposes of these methods are the preservation of the whole content of the MS image and the addition of further information obtained from the Pan image, through spatial filtering.

In the seminal paper by Tu *et al.* [252], the mathematical development that led to the fast algorithm of GIHS was also applied to the wavelet fusion described by the flowchart in Figure 7.3. The main result was that only the equivalent lowpass filter that generates the baseband image (see leftmost plot in Figure 5.4) affects the pansharpened image. Hence, the contribution of the Pan image to the fused product is achieved by calculating the difference between **P** and its lowpass version $\mathbf{P}_L$. This strictly holds for UDWT; there is a slight difference in the case of DWT because of aliasing introduced by decimation. This issue will be addressed in Section 9.3.2. Thus, the most general MRA-based fusion model may be stated as:

$$\hat{\mathbf{M}}_k = \tilde{\mathbf{M}}_k + g_k \left(\mathbf{P} - \mathbf{P}_L \right), \ \forall k = 1, \ldots, K \tag{7.9}$$

According to (7.9) the different approaches and methods belonging to this class are uniquely characterized by the lowpass filter employed for obtaining the image $\mathbf{P}_L$ and by the set of injection gains $\{g_k\}_{k=1,\ldots,K}$. The flowchart for the fusion procedure is outlined in Figure 7.4.

The mathematical expression of $\mathbf{g}$ assumes different forms in the literature. Among the possible types of filters, the simplest scheme is achievable by using the *box* mask (that is, a mask with uniform weights, implementing an average filter) and additive injection, which leads to the pansharpening algorithm known as *Highpass Filtering* (HPF) method [66, 226, 227].

However, setting $g_k = 1$, $\forall k$, causes miscalibration of Pan on MS, unless the former has been preliminarily cross-calibrated on each of the MS bands, that is, $\mathbf{P} = [\text{Pan} - \mu(\text{Pan})] \cdot \frac{\sigma(M_k)}{\sigma(P_L)} + \mu(M_k)$ [258]. That is the main drawback encountered by HPF.

Of greater interest is the *Highpass Modulation* (HPM) method [226], reinterpreted in [174] within the context of radiative transfer model and renamed *Smoothing Filter-Based Intensity Modulation* (SFIM),

$$\hat{\mathbf{M}}_k = \tilde{\mathbf{M}}_k + \frac{\tilde{\mathbf{M}}_k}{\mathbf{P}_L} \left(\mathbf{P} - \mathbf{P}_L\right), \ \forall k = 1, \ldots, K, \tag{7.10}$$

in which the details injected into the kth band are weighted by the ratio of the interpolated MS and the lowpass Pan $\mathbf{P}_L$, with the aim of reproducing the local intensity contrast of Pan. Furthermore, if a unique lowpass image $\mathbf{P}_L$ is used for all the MS bands (this may not occur if a different filter is used for each band and/or if the Pan image has been histogram matched to each of the MS bands), this algorithm clamps the spectral distortion of $\hat{\mathbf{MS}}$ to that of $\tilde{\mathbf{MS}}$, which can be quantified by SAM, and thus is an element of the *Spectral Distortion Minimization* (SDM) class [22, 23, 13].

7.3.4 Optimization of MRA Based on Instrument MTF

In this section we will review the main results of Aiazzi *et al.* [14], in which evidence is given that optimization of the unique lowpass filter used for MRA-based fusion can be achieved if its response in the spatial frequencies matches the MTF of the imaging instrument.

Figure 2.4(a) shows the theoretical MTF of an imaging system. As a first approximation, the MTF is equal to the modulus of the Fourier transform of the optical point spread function (PSF) of the imaging system. In principle, two spectral replicas originated by 2-D sampling of the radiance signal with the same sampling frequency along- and across-track, should cross each other at the Nyquist frequency (half of the sampling frequency) with magnitude values equal to 0.5. However, the scarce selectivity of the response prevents from using a sampling frequency that yields a Nyquist frequency with magnitude equal to 0.5. As a trade-off between (maximum) spatial resolution and (minimum) aliasing of the sampled signal, the sampling frequency is usually chosen such that the corresponding magnitude at the Nyquist frequency is comprised between 0.2 and 0.3.

This situation is depicted in Figure 2.4(b) portraying the true MTF of an MS channel, which also depends on the platform motion (narrower along track), on atmospheric effects and on the finite sampling. A different situation occurs for the MTF of the Pan channel, whose extent is mainly dictated by diffraction limits (at least for instruments with a resolution around 1 m). In this case, the cutoff magnitude may be even lower (e.g., 0.1) and the appearance of the acquired images is rather blurred. However, whereas the enhancing Pan image that is commonly available has been already processed for MTF restoration, the MS bands cannot be preprocessed analogously because of SNR constraints. In fact, restoration implies a kind of inverse filtering, that has the effect of increasing the noisiness of the data.

Ultimately, the problem may be stated in the following terms: an MS band resampled at the finer scale of the Pan image lacks high-spatial frequency components, that may be inferred from the Pan image via a suitable interscale injection model. If the highpass filter used to extract such frequency components from the Pan image is taken so as to approximate the complement of the MTF of the MS band to be enhanced, the high-frequency components that have been damped by the MTF of the instrument, can be restored. Otherwise, if spatial details are extracted from the Pan image by using a filter having normalized frequency cutoff at exactly the scale ratio between Pan and MS (e.g., 1/4 for 1 m Pan and 4 m MS), such frequency components will not be injected. This occurs with critically subsampled wavelet decompositions, whose filters are constrained to cutoff at exactly an integer fraction (usually a power of two) of the Nyquist frequency of Pan data, corresponding to the scale ratio between Pan and MS.

An attractive characteristic of the redundant pyramid and wavelet decompositions proposed by the authors [12, 112] is that the lowpass reduction filter used to analyze the Pan image may be easily designed such that it matches the MTF of the band into which the details extracted will be injected. Figure 7.5(a) shows examples for three values of magnitude cutoff. The resulting benefit is that the restoration of spatial frequency content of the MS bands is provided by the MRA of Pan through the injection model. Such a facility is not immediate with conventional DWT and UDWT decompositions [107].

A forerunner of this rationale was the work by Núñez *et al.* [196], even if no considerations on MTF were made. Figure 7.5(b) shows that the Gaussian-like frequency response of the cubic-spline wavelet filter used to generate ATW matches the shape of the MTF of a typical V-NIR band, with a cutoff magnitude value equal to 0.185. The complementary highpass filter, yielding the detail level to be injected for 1:4 fusion, retains a greater amount of spatial frequency components than an ideal filter, such as that used by the standard GLP and ATW [12, 112], thereby resulting in a greater spatial enhancement.

As a matter of fact, also quality assessment benefits from the MTF of the MS instrument. The MTF, a lowpass filter, and its complement, a highpass fil-

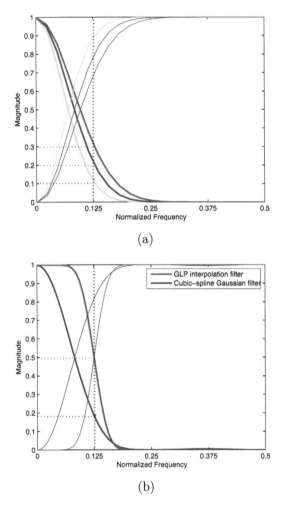

FIGURE 7.5: 1-D frequency responses of equivalent filters for separable 2-D 1:4 multiresolution analysis (lowpass and highpass filters generate approximation and detail, respectively). (a): sample MTF-adjusted Gaussian-shaped filters with magnitudes of 0.1, 0.2, and 0.3 at the Nyquist frequency; (b): frequency selective 11-tap interpolation filter (red) and nonselective cubic-spline filter for ATW generation (blue).

ter, define the frequency channels, in which consistency measurements of pansharpened MS data to its components, original MS and original Pan, should be made [147], respectively, according to Khan's protocol (Section 3.2.2.4).

7.3.4.1 ATW

Starting from Núñez *et al.* [196], the undecimated "à-trous" analysis soon emerged as a very effective MRA for image fusion [125]. Indeed, even if non-orthogonality (which implies that a wavelet plane could retain information for a neighboring plane) could compromise the spectral quality of the fused product [125], its beneficial characteristics such as the shift-invariance property [256] and the capability of being easily matched to the sensor MTF [14] is valuable for producing accurate pansharpened image.

The most widely used implementation of the "à-trous" filter based on the sequential application (thanks to the separability property), in the vertical and horizontal direction, of the 1-D kernel [243]

$$h = \begin{bmatrix} 1 & 4 & 6 & 4 & 1 \end{bmatrix}, \tag{7.11}$$

derived from the choice of a B_3 cubic spline as the scaling function of MRA [244]. The frequency responses of the prototype filter and of the equivalent bandpass filters are shown in Figure 5.9.

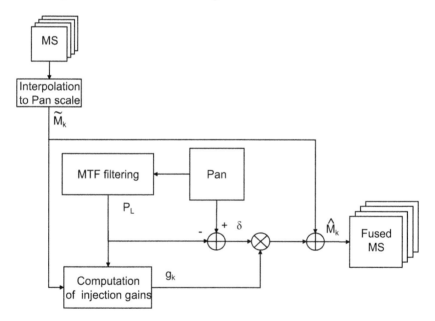

FIGURE 7.6: Flowchart of MRA-based pansharpening exploiting ATW.

7.3.4.2 GLP

The resolution reduction needs to obtain the lowpass Pan image, $\mathbf{P}_L$, at the original MS scale can be performed in one or more steps, namely by employing a single lowpass filter with a cutoff frequency equal to $1/r$ and decimating

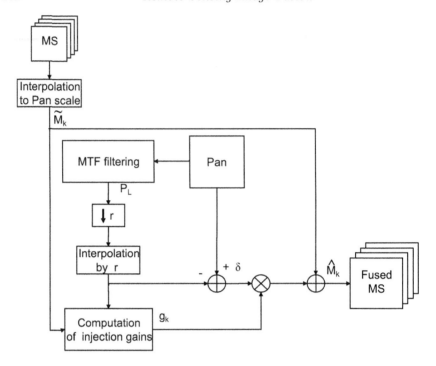

FIGURE 7.7: Flowchart of MRA-based pansharpening exploiting GLP.

by r, or by multiple fractional steps. The latter, including the former as a particular case, is commonly referred to as pyramidal decomposition and dates back to the seminal work by Burt and Adelson [54], which utilized Gaussian lowpass filters to carry out the analysis steps. The corresponding differential representation, achieved by calculating the differences between the Gaussian pyramid levels, is named *Laplacian pyramid* and has later been proven to be invaluable for pansharpening purposes [12].

Figure 7.7 illustrates the difference between ATW and LP, or better GLP, which became the synonym of LP tailored to fusion, even though the scale ratio is not fractional. The presence of the downsampler and of the interpolator in the flowchart makes the GLP exhibit a negligible difference against the ATW if the aliasing is weak, that is, filters are selective or their magnitude at the Nyquist frequency is small, and interpolation is close to being ideal, that is, at least bicubic. Otherwise, GLP fusion may somewhat differ from ATW fusion, even if both use the same filter. A detailed analysis of the difference between ATW and GLP fusion will be the object of Section 9.3.2.

Gaussian filters can be tuned to closely match the MTF of the sensor. This allows the details that are not seen by the MS sensor due its the coarser spatial resolution to be extracted from the Pan image. Since the Gaussian mask is defined by a single parameter (that is, its standard deviation), its frequency response is fully specified by fixing it. To this aim, the value of the amplitude response at the Nyquist frequency is used, being that it is commonly provided by the manufacturer as a sensor specification, or obtained through on-orbit measurements. However it is useful to keep in mind that, components aging can induce a significant incertitude of this parameter, also in the more favorable case of on-orbit estimation.

As a further example of GLP fusion exploiting an analysis filter matched with the MS sensor MTF, we consider an algorithm that relies upon the optimization of the injection coefficients by least square fitting. In particular, it is based on (7.9), in which the coefficients are calculated as:

$$\mathbf{G}_k = \frac{\mathrm{cov}(\tilde{\mathbf{M}}_k, \mathbf{P}_L)}{\mathrm{var}(\mathbf{P}_L)}, \tag{7.12}$$

where, in general, $\mathbf{P}_L$ depends on the kth band. This injection rule, identical to that coming out from the mathematical development of GS [26], leads to an effective method known as GLP with a context-based decision (GLP-CBD) [12, 14], since the injection coefficient can be locally optimized in non-overlapped blocks. The consequent performance improvement is paid in terms of a significant increase in execution time.

7.3.4.3   MRA-optimized CS

Figure 7.2 shows that there is a lowpass filter in the optimized CS scheme. Such a filter has the purpose of equalizing the spatial frequency content of the MS and Pan images before the multivariate regression aimed at finding the optimal spectral weights is carried out. In practice, the Pan image is smoothed by means of a digital filter simulating the process undergone by the continuous MS bands before they are sampled. What filter would be better suited than the one matching the average MTF of the spectral channels? Average because there is a unique filter for Pan and the MTFs of the spectral channels are slightly different from one another, since the resolution of the optics decreases as the wavelength increases.

Figure 7.8 shows the upgraded flowchart, in which MTF filtering is used before the Pan image is downsampled. Besides spectral weights (CS) and spatial filters (MRA), the injection model described by the g_ks can be defined in order to optimize the quality of fusion products [20], according to some index, such as Q4 [113, 114] or QNR [146, 148]. To this purpose, also adaptive, that is, data dependent, space-varying injection coefficients can be defined [25]. Adaptive injection models may be calculated on local sliding windows, or even better on block partitions of the image and then interpolated over all image pixels, to save computation.

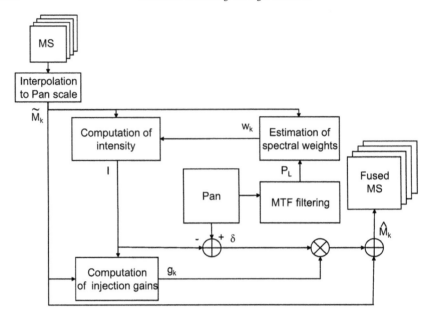

FIGURE 7.8: Optimized CS pansharpening scheme exploiting MTF.

7.3.5   Hybrid Methods

Hybrid methods are those in which MRA is coupled with a spectral transform, like in CS methods. There is plenty of such methods because the number of combinations between transforms (IHS, PCA, GS) and types of MRA (DWT, UDWT, ATW, GLP, complex wavelets, curvelets, contourlets, bandelets, ridgelets [86], etc.) is high [71] [173].

All the previous wavelet methods are based on the choice of unitary injection coefficients $\{g_k\}_{k=1,\ldots,K}$. However, some further improvements can be achieved by injecting the details using the HPM paradigm (7.10) [258]. As an example of MRA-based method following the modulation approach, we recall the *Additive Wavelet Luminance Proportional* (AWLP) [200] that uses the more general fusion formula reported in (7.9), with the space-varying injection coefficients defined as

$$\mathbf{G}_k = \frac{\tilde{\mathbf{M}}_k}{\frac{1}{K}\sum_{i=1}^{K}\tilde{\mathbf{M}}_i}, \quad \forall k = 1,\ldots,K. \tag{7.13}$$

Besides IHS, PCA has also been coupled with DWT and UDWT [126]. In addition, adaptive PCA has been combined with contourlet decomposition [229]. By following the guidelines of AWL and later of AWLP within the framework presented in this book, a spectral transformation coupled with a certain type of MRA is nothing but a flowchart like the one shown in Figure 7.4, possibly with a specific injection model.

7.4   Simulation Results and Discussion

In this section, fusion simulations are devoted to assess the quality, both visually and objectively, of the pansharpened images provided by the fusion methods presented in the previous sections. For this purpose, images acquired by VHR sensors, as IKONOS and QuickBird are considered both at degraded and full scale to compute quality and distortion indices and obtain a comprehensive objective assessment.

7.4.1   Fusion of IKONOS Data

In the following, we present quality assessments performed on very high-resolution image data acquired by the spaceborne IKONOS MS scanner, on the city of Toulouse, France. The four MS bands of IKONOS span the visible and NIR wavelengths and are non-overlapped, with the exception of B1 and B2. The bandwidth of Pan embraces the interval 450÷950 nm. The dataset has been geocoded to 4 m (MS) and 1 m (Pan) pixel size. A 512 × 512 portion of the original Pan image is reported in Figure 7.9(a) and the corresponding original MS image resampled to the scale of Pan is shown in Figure 7.9(b). All statistics have been calculated on the whole fused image, unless explicitly stated.

The following fusion methods have been compared: Generalized Intensity-Hue-Saturation (GIHS) [252]; Gram-Schmidt spectral sharpening [157], either in Mode 1 (GS1) or Mode 2 (GS2), with enhanced spectral transformation [26] (GSA) and context-adaptive detail injection model [25] (GSA-CA); GLP-based method [12] with MTF adjustment [14] either without (GLP) or with context-adaptive injection model [25] (GLP-CA). The "void" fusion method, corresponding to plain (bicubic) resampling of the MS dataset at the scale of Pan is included in the comparisons and referred to as EXP. It is to be assumed as the reference for evaluating color distortions.

First, the MS and Pan images were downsized by four to allow quantitative evaluations to be performed according to Wald's synthesis property. A Gaussian-like lowpass prefilter with amplitude equal to 0.25 at Nyquist frequency has been used for all bands to avoid aliasing.

Figure 7.10(a) shows the result of GS1, that is, the synthetic low-resolution Pan is obtained as pixel average of the four MS bands. The spatial enhancement is impressive; however, some color distortion appears on the green areas on the left of the river and on the river itself. The green areas are too bright while the color of the river is too dark. These spectral distortions are not present in Figure 7.10(b) that shows the outcome of GS2, that is, the synthetic Pan is obtained as a lowpass-filtered version of the full-resolution Pan. As a counterpart, Figure 7.10(b) is less sharp than Figure 7.10(a).

Both Figures 7.10(c) and 7.10(d) result from the generalized versions of

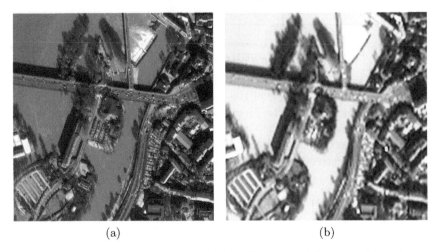

(a)   (b)

FIGURE 7.9: Original 512 × 512 IKONOS images. (a): original Pan image; (b): original MS bands (4 m) resampled to the scale of Pan image (1 m).

the GS algorithm. Figure 7.10(c) has been obtained by synthesizing the low-resolution Pan as a weighted sum of all spectral bands (GSF), according to the fixed weights proposed by Tu *et al.* [251] ($\mathbf{w} = \{1/12, 1/4, 1/3, 1/3\}$). Figure 7.10(c) appears as sharp as Figure 7.10(a), but the spectral distortion is mitigated. The result obtained with the adaptive preprocessing (GSA) is reported in Figure 7.10(d). The weights calculated on the whole image are reported in the first row of Table 7.2. The spectral fidelity with respect to the original MS data of Figure 7.9(b) is impressive, while spatial details appear as sharp as those of Figure 7.10(a). The second row of Table 7.2 reports the weights computed through multivariate regression between the original MS bands and the reduced Pan of the QuickBird image shown in Figure 7.14. Though the spectral responses of IKONOS (Figure 2.12) and QuickBird (Figure 2.13) are quite similar, the weights in Table 7.2 are significantly different for the two sensors, since they are strongly dependent on the landscape and not only on the imaging instrument.

Figure 7.11(b) reports the result obtained by applying GIHS while Figure 7.11(c) refers to GIHSF. Both methods have been implemented as reported in Tu *et al.* [251]. Spatial details are very sharp but spectral distortions are evident in both images (the original MS image is also shown in Figure 7.11(a) to ease visual comparisons). Finally, Figure 7.11(d) shows the result obtained by the GIHSA method showing a very high-spectral fidelity which is comparable to that provided by GS2 and GSA.

Small 128×128 details produced by all the methods, extracted from the images reported in Figure 7.9, are displayed in Figure 7.12. Spectral distortions are noticeable as changes in color hue of the river with respect to the resampled low-resolution original MS of Figure 7.12(b). The images of Fig-

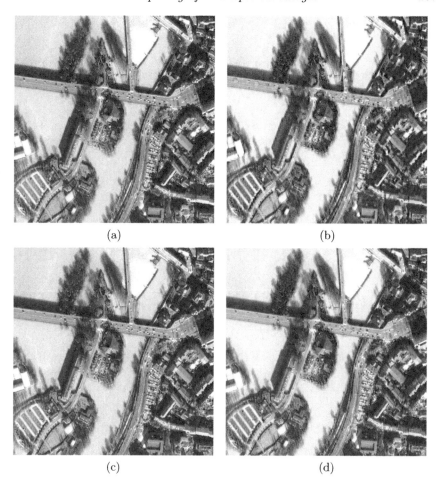

(a)   (b)

(c)   (d)

FIGURE 7.10: Examples of full-scale spatial enhancement of GS fusion algorithms displayed as 512×512 true-color compositions at 1 m pixel spacing for the IKONOS image. (a): GS-mode 1 fusion (GS1); (b): GS-mode 2 (GS2); (c): GS with fixed weights (GSF); (d): GS with adaptive weights (GSA).

TABLE 7.2: Optimal regression-based weights computed between original MS bands and reduced Pan for IKONOS and QuickBird images.

	$\hat{w}_1$	$\hat{w}_2$	$\hat{w}_3$	$\hat{w}_4$
IKONOS	0.067	0.189	0.228	0.319
QuickBird	0.035	0.216	0.387	0.412

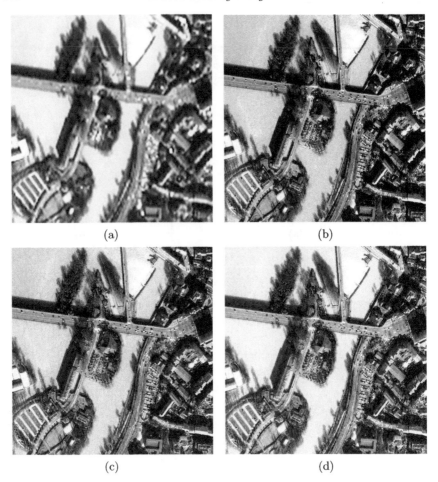

(a)   (b)

(c)   (d)

FIGURE 7.11: Examples of full-scale spatial enhancement of GIHS fusion algorithms displayed as 512×512 true-color compositions at 1 m pixel spacing for the IKONOS image. (a): original MS bands (4 m) resampled to the scale of Pan image (1 m); (b): generalized IHS with equal weights (GIHS); (c): generalized IHS with unequal fixed weights (GIHSF); (d): generalized IHS with adaptive weights (GIHSA).

ures 7.12(f) and (i), GSA and GIHSA, respectively, exhibit high *spectral* and *spatial* quality. All fused images show impressive spatial details since are all derived from CS methods. Spatial distortions are noticeable only in Figure 7.12(d) (GS2) especially for the cars on the bridge. However, GS2 *de facto* belongs to the MRA methods, because the synthetic intensity component is obtained by lowpass filtering and decimating the original Pan image.

Visual judgement is corroborated by quantitative assessments. The score values are reported in Table 7.3 and show that the regression-based enhance-

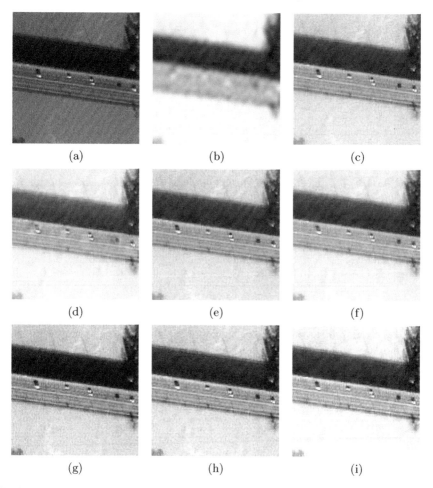

FIGURE 7.12: IKONOS: Results of fusion algorithms displayed as 128 ×
128 true-color compositions at 1 m SSI. (a): original Pan image; (b): original
MS bands (4 m) resampled to the scale of Pan image (1 m); (c): GS1; (d):
GS2; (e): GSF; (f): GSA; (g): GIHS; (h): GIHSF; (i): GIHSA.

ment strategy improves both GS and GIHS fusion algorithms with respect
to all other nonadaptive implementations. In particular, GSA results better
than GIHSA when considering Q4 and SAM, while GIHSA is superior to GSA
regarding ERGAS. Since GIHSA and GSA adopt the same **I** component, this
is evidently due to the different $\{g_k\}$ gains adopted by the two algorithms. For
the IKONOS test image GIHSA weights are more efficient than those of GSA.
The only slight exception in Table 7.3 is represented by SAM of GIHSF and
GIHSA. This is however compensated by the significant improvement of ER-
GAS. Actually, the MSE minimization, in this case, has produced regression

TABLE 7.3: IKONOS: Average cumulative quality indices between original 4 m MS and fused images obtained from 16 m MS and 4 m Pan.

4:1	EXP	GS1	GS2	GSF	GSA	GIHS	GIHSF	GIHSA
Q4	0.630	0.857	0.849	0.860	0.864	0.746	0.834	0.850
SAM (°)	4.85	4.19	4.03	4.06	3.82	4.17	3.92	3.94
ERGAS	5.94	3.83	3.81	3.71	3.55	4.19	3.38	3.23

TABLE 7.4: IKONOS: Average cumulative quality indices between original 4 m MS and fused images on urban areas.

4:1	EXP	GS1	GS2	GSF	GSA	GIHS	GIHSF	GIHSA
Q4	0.617	0.866	0.873	0.866	0.870	0.830	0.811	0.839
SAM (°)	5.59	4.51	4.55	4.51	4.43	5.15	5.12	5.17
ERGAS	5.46	3.26	3.20	3.26	3.18	3.56	3.60	3.44

TABLE 7.5: IKONOS: Average cumulative quality indices between original 4 m MS and fused images on vegetated areas.

4:1	EXP	GS1	GS2	GSF	GSA	GIHS	GIHSF	GIHSA
Q4	0.645	0.793	0.817	0.793	0.844	0.785	0.760	0.805
SAM (°)	3.22	3.04	3.08	3.04	2.72	3.08	3.07	3.07
ERGAS	3.14	2.31	2.27	2.31	2.00	2.14	2.16	2.03

TABLE 7.6: IKONOS: Average cumulative quality indices between original 4 m MS and fused images on homogeneous areas.

4:1	EXP	GS1	GS2	GSF	GSA	GIHS	GIHSF	GIHSA
Q4	0.876	0.689	0.869	0.687	0.860	0.662	0.665	0.910
SAM (°)	1.03	3.64	1.24	3.64	1.26	2.03	1.81	1.01
ERGAS	1.23	5.15	1.15	5.15	1.17	3.73	3.16	0.90

coefficients that are not the best for SAM. A nonnegligible improvement of SAM of GIHSA may be obtained by just slightly decreasing $\hat{w}_1$ and increasing correspondingly $\hat{w}_2$. As expected, the improvement of SAM is paid in terms of ERGAS.

Tables 7.4, 7.5, and 7.6 report the quality figures that have been evaluated on the three classes of urban, vegetated, and homogeneous areas, respectively.

As expected, figures depend on the quantity of details that are present on the landscape. Urban class figures are the worst while those of a homogeneous class are the best. GSA and GIHSA are always better than their nonadaptive counterparts with only a few minor exceptions; also in the other cases their score is quite near to the best one. In any case, their ERGAS is always the lowest. Another general consideration is that the fixed weights defined in Tu *et al.* [251] do not perform better than the baseline in which all the weights are equal. This is a firm indication that fixed coefficients are not suitable to cope with such variabilities as those introduced by scene variations, atmospheric conditions, and attitude or aging of the acquisition platform. Conversely, regression strategy guarantees a greater flexibility and more steady performances. The superiority of GSA and GIHSA is evident on vegetated areas, on which they largely outperform the other schemes, demonstrating that modeling of spectral response is effective on vegetation. The performance of constant weight methods on homogeneous areas (mainly extracted from the river appearing in Figure 7.9) are disappointing. Since homogeneous areas are dealt with, the only explanation is that they are not able to efficiently model the local average value of the synthetic intensity image. The ERGAS values are particularly low for GSA, GS2, and GIHSA demonstrating their effectiveness.

The IKONOS data have been pansharpened at the full 1 m scale and QNR protocol has been applied for objective quality assessment. Table 7.7 reports values of the spectral distortion D_λ and of the spatial distortion D_s that are separately calculated at full scale. Both distortions can be merged into a unique normalized quality index, referred to as QNR, for an easy comparison with indices like UIQI and Q4.

According to the QNR entry in Table 7.7, the ranking of fusion methods is substantially analogous to that may be obtained at reduced scale according the Q4 index. A notable exception is the high QNR value of the resampled image (EXP), which is due to the null spectral distortion measured when no detail injection is performed. A proper nonlinear combination of D_λ and D_s, as that proposed in Alparone *et al.* [32], might exactly match the trend of Q4, because the dependence of Q4 on spectral and spatial distortions is implicit and unknown.

TABLE 7.7: Quality measurements of pansharpened IKONOS data at full scale (1 m). QNR is defined as $(1 - D_\lambda)(1 - D_s)$ and is a global quality measurement in [0,1], analogous to Q4.

	EXP	GIHS	GS	GSA	GSA-CA	GLP	GLP-CA
D_λ	0.000	0.125	0.055	0.056	0.034	0.075	0.066
D_s	0.168	0.139	0.096	0.087	0.073	0.099	0.082
QNR	0.832	0.753	0.854	0.862	0.895	0.834	0.857

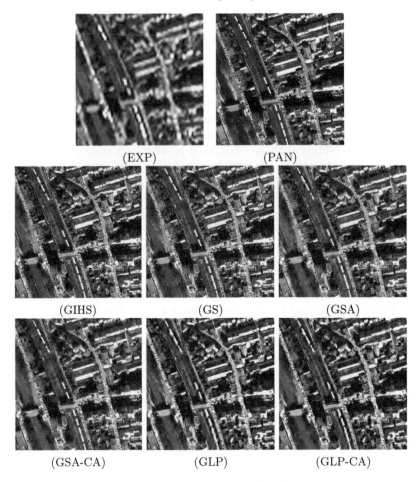

(EXP)   (PAN)

(GIHS)   (GS)   (GSA)

(GSA-CA)   (GLP)   (GLP-CA)

FIGURE 7.13: True-color details at full scale (1 m): bicubically interpolated MS, Pan, and six fusion products.

Small fragments of original and pansharpened images are shown in Figure 7.13 for all methods, including the *void* fusion EXP. By watching the EXP and GIHS icons, one can realize why the QNR of EXP is higher than that of GIHS: because the former is underenhanced ($D_s = 0.168$), but the latter is overenhanced ($D_s = 0.139$) and also exhibits mediocre spectral quality ($D_\lambda = 0.125$) compared to the ideal spectral quality of EXP ($D_\lambda = 0$).

7.4.2   Fusion of QuickBird Data

MRA fusion algorithms have been also assessed on QuickBird data. Very high-resolution images collected on June 23, 2002, at 10:25:59 GMT+2 on the urban area of Pavia, Italy, have been considered for simulations. All data are geocoded and resampled with an SSI of 2.8 m and 0.7 m for MS and Pan, respectively. Following the synthesis property of Wald's protocol (see Section 3.2.2.1), the MS and Pan data have been spatially degraded by 4 (to yield 2.8 m Pan and 11.2 m MS) and used to synthesize the MS bands at 2.8 m. Thus, the true MS data at 2.8 m are available for objective distortion measurements. We wish to remind that simulations on spatially degraded data is aimed at adjusting the fusion algorithm in such a way that once the best objective fidelity scores with the reference originals have been obtained, the same algorithm will run on the true higher-resolution data and produce the best results. Rather than trying to obtain the best absolute scores, it is important to measure how much a fusion method is capable of improving the quality of the fused product with respect to that of the resampled MS data that constitute its starting point. MTF-matched Gaussian filters have been used for prefiltering the MS data before decimation. Global scores, like SAM, ERGAS, and Q4 have been used for assessments.

A comparison of GLP-based methods with SDM and CBD injection models was carried out with the baseline MRA-based method, that is, HPF [66], with a 7 × 7 box filter for obtaining the best performance. The SDM and CBD injection models are applied to GLP, which is achieved with MTF-matched filters. The parameters in Table 7.8 measuring the global distortion of pixel vectors, either radiometric (ERGAS, which should be as low as possible) or spectral (SAM, which should be as low as possible), and both radiometric and spectral (Q4, which should be as close as possible to one) provide a comprehensive measurement of fusion performance. EXP denotes the case in which the degraded MS data are resampled through the 23-tap expansion filter of GLP [12] and no injection of details is made. The SAM attained by MTF-GLP-CBD is lower than that of MTF-GLP-SDM (identical to that of EXP), thanks to the *spectral unmixing* capabilities of the former compared to the latter.

TABLE 7.8: Average cumulative quality indices between 2.8 m MS spectral vectors and those obtained from fusion of 11.2 m MS with 2.8 m Pan. MS bands degraded with MTF-matched filters.

	Q4	SAM (deg.)	ERGAS
EXP	0.641	2.46°	1.980
MTF-GLP-SDM	0.819	2.46°	1.608
MTF-GLP-CBD	0.834	2.13°	1.589
HPF (7 × 7)	0.781	2.81°	2.132

Figure 7.14 displays true-color compositions of the resampled 2.8 m MS bands and of the spatially enhanced bands, all at 0.7 m. True-color visualization has been chosen intentionally, because pansharpening of MS bands

falling partly outside the bandwidth of Pan, as in the case of the blue band, is particularly critical.

A visual inspection highlights that all the spectral signatures of the original MS data are carefully incorporated into the sharpened bands. Thanks to the two injection models, the texture of the canopies, which is highlighted by the Pan image, but mostly derives from the NIR band, which is outside the visible wavelengths, appears to be damped in the SDM and CBD fusion products. HPF is geometrically rich and detailed, but overenhanced, especially on vegetated areas. A visual analysis carried out, for example, on the small circular square surrounded by trees, reveals that the methods exploiting MTF-tailored MRA yield sharper and cleaner geometrical structures than the method that does not exploit an MTF-matched digital filter.

7.4.3 Discussion

The results of simulations give evidence that the good visual appearance and the spectral content preservation represent the main salient features of the CS and MRA methods, respectively. Accordingly, approaches of the first class aimed at improving the spectral quality and those of the second class properly designed for enhancing the spatial properties obtain the best results. Indeed, very interesting performances are attained by *adaptive* CS approaches (in particular on four band datasets), with the reduction of the spectral distortion, and by some MRA algorithms, which benefit from an instrument-matched detail-extraction. In particular, the match of the lowpass filter with the sensor's MTF allows the unsatisfactory spatial enhancement of classical MRA fusion products to be significantly improved.

From a computational point of view, CS-based methods, at least in their nonadaptive versions, are faster than MRA-based methods, as the spatial filtering steps significantly slows down computation.

The analysis carried out in this work allowed us to confirm some features of the validation procedures: The reduced scale protocol leads to a very accurate evaluation of the quality indices, but the scale invariance hypothesis is hardly verified in practice. Furthermore, the operation for generating a synthetic version of the MS image at reduced scale introduces a strong bias in the analyzed algorithms, privileging those employing a similar procedure for extracting the spatial details. On the contrary, the full scale validation gets quite accurate results with respect to algorithms of the same family, whereas being less reliable when the comparison includes algorithms belonging to different classes. This validation approach, which exploits the QNR index, has the advantage of validating the products at the original scale, thus avoiding any hypothesis on the behavior at different scales. However, due to the lower precision of quality indices (since a reference is missing), the results of this analysis can be affected by some mismatches between the quantitative results and the visual appearance of the fused image, for example, as in the GIHS method.

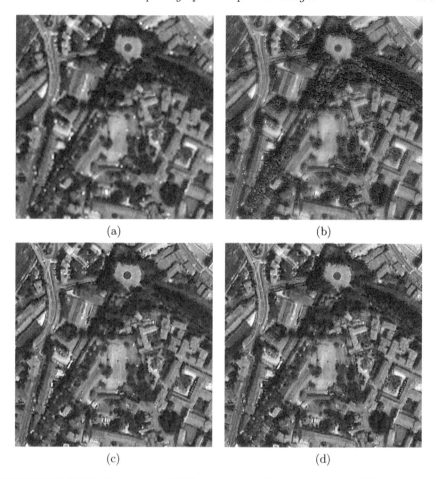

(a)   (b)

(c)   (d)

FIGURE 7.14: Examples of full-scale spatial enhancement of fusion algorithms displayed as 512 × 512 true-color compositions (B3, B2, and B1 as R-G-B channels) at 0.7 m pixel spacing. (a): original MS bands (2.8 m) resampled at the scale of the Pan image (0.7 m); (b): HPF method, with 7× 7 box filter; (c): GLP-CBD with MTF-adjusted filter; (d): GLP-SDM with MTF-adjusted filter.

7.5 Concluding Remarks

In this chapter a critical review and categorization of state-of-the-art pansharpening methods has been attempted. The main contribution is defining a computational framework in which the two main categories of methods, based on CS and MRA, respectively, as well as CS+MRA hybrid versions, may be accommodated and parametrically optimized. Specifically, CS-based methods are spectral methods and are optimized through a spectral matching of the MS and Pan channels, achieved by matching the Pan image to a linear combination of the MS bands. MRA-based methods are spatial methods and are optimized by matching the equivalent digital lowpass filter to the MTF of the imaging system. Some widely used algorithms belonging to these families, in both optimized and nonoptimized versions, have been comparatively evaluated by simulation tests.

The particular characteristics of the two classes of pansharpening algorithms have been evidenced during the assessment phase. Specifically, the appealing features of the CS-based methods from a visual point of view has been easily shown by highlighting the absence of aliasing artifacts. The latter property, together with the robustness of these methods against the errors induced by the misregistration of the available images, has supported their large use also featured by a low computational burden. On the other side, the best overall performances are often achieved by MRA-based methods, which are characterized by a more exact reproduction of the MS spectral characteristics and that can be easily designed to match the MS sensor properties in the phase of spatial details extraction. Furthermore, this class of algorithms is currently drawing increasing attention, thanks to their temporal coherence, they can be employed for multiplatform data fusion. This kind of robustness turns out to be very helpful when the fusing images are acquired with a time delay, as, for example, when they are provided by sensors mounted on different RS platforms. Chapter 9 is entirely devoted to highlight, both theoretically and through simulations, the substantial and complementary differences between CS and MRA fusion methods in the presence of aliasing in the MS data and registration inaccuracies, as well as of not simultaneous acquisitions of MS and Pan.

Chapter 8

Pansharpening of Hyperspectral Images

8.1   Introduction

This chapter is devoted to the problem of extending pansharpening or more generally image fusion methods to hyperspectral (HS) image data. There are a number of problems that make such an extension crucial, or at least nontrivial. The most relevant issues will be addressed in Section 8.2. After a concise literature survey in Section 8.3, two methods recognized as promising will be reviewed. Simulations and comparisons with the state of the art will be presented in Section 8.4.

Since HS technology has become mature, remote sensing platforms can be developed with the capability of concurrently capturing HS images along with panchromatic (Pan) or multispectral (MS) images of higher-spatial resolutions. The difference in spatial resolutions is a consequence of the fundamental trade-off in the design of electro-optical systems between spatial resolution, spectral resolution, and radiometric sensitivity. Thus, HS resolution enhancement, or HS sharpening, refers to the joint processing of the data in order to synthesize an HS image product that ideally exhibits the spectral characteristics of the observed HS image at the spatial resolution, and hence sampling, of the higher-resolution image.

An operative example is the Earth Observer 1 (EO-1) satellite launched in 2001 and hosting the Advanced Land Imager (ALI) MS scanner and the

Hyperion imaging spectrometer. ALI acquires nine MS channels and a Pan image with an SSI of 30 m and 10 m, respectively. Hyperion is an HS imager capable of observing 220 spectral bands (0.4 to 2.5 nm) with 30 m SSI. The swaths of ALI and Hyperion are partially overlapped in such a way that there is a part of the scene, in which Pan, MS and HS data are simultaneously available. On this scene HS fusion algorithms can be developed and tested for a 3:1 scale ratio.

The upcoming PRISMA satellite system, operated by Italian Space Agency (ASI), will make simultaneously acquired HS and Pan data of the same scene available to the user community. The Italian acronym stands for *hyperspectral precursor of application mission*. The payload of PRISMA is constituted by two imaging spectrometers and a Pan camera. The spectrometers operate in the V-NIR [400–1010 nm] and SWIR [920–2500 nm] spectral ranges, respectively, with a nominal spectral sampling interval of 10 nm; their SSI is 30 m with a 12 bit radiometric resolution. The SSI of the Pan image is 5 m and its spectral range is [400–700 nm]. Image fusion can be designed in such a way that an HS data cube can be produced with an SSI of 5 m and a spectral resolution of 10 nm.

Also, the *Environmental Monitoring and Analysis Program* (EnMAP) satellite operated by the German Aerospace Center (DLR) will provide HS data cubes with the same nominal SSI and spectral resolution as PRISMA. Unfortunately, simultaneous acquisition of HS data with higher-resolution imagery was not scheduled for the mission, thus preventing a straightforward image fusion from being achieved. In this case, HS pansharpening algorithms must work with a Pan image acquired by another platform and thus shifted in time and possibly taken with a different angle.

New perspectives for HS research and for pansharpening will be opened by such missions as SHALOM (*Spaceborne Hyperspectral Applicative Land and Ocean Mission*), a joint Earth observation program of Italian and Israeli Space Agencies. SHALOM shall collect HS data with a 10 m SSI, a swath 10 Km large, a [400–2500 nm] spectral range, and a 10 nm nominal spectral resolution. A Pan image with 2.5 m SSI will be also acquired on the same scene, motivated by the requirement of pansharpening. The high-spatial resolution of SHALOM is obtained by a steering capability of the platform that allows the same portion of the Earth surface to be observed for a longer acquisition time through Earth motion compensation. The increment in the integration time is beneficial to improve the radiometric performances with a smaller instrument aperture and thus with a lower platform weight.

8.2 Multispectral to Hyperspectral Pansharpening

In the previous section, we noted that the different spatial resolution existing between sensors is a consequence of the trade-off in the design of electro-optical systems. Such a trade-off is aimed at balancing some aspects the main of which are SNR, physical dimensions of sensor arrays, and transmission rate from the satellite to the receiving stations.

SNR usually decreases when spatial and spectral resolutions increase. Notwithstanding the modern *Time Delay Integration* (TDI) technology is the key to increase the signal-to-noise ratio (SNR) for high-resolution Pan and MS imagery, such a solution is not feasible for HS instruments, because the two-dimensional array that is placed on the focal plane of the instrument and acquires in the across-track (x) and spectral (λ) directions should become three-dimensional, that is, x, λ and the along-track coordinate y, in which delayed versions of the x-λ plane are acquired. So, especially for spaceborne instruments we expect that the spatial resolution of HS sensors will remain limited to tenths of meters, unlike a steering strategy, that is, tilting the platform during acquisition, as in the case of SHALOM, is adopted.

Physical dimensions of sensor arrays and thus of sensor elements are also crucial. They cannot be reduced without impairing the SNR. Also, technological developments in solid state devices will allow more efficient detectors with better SNR to be developed, spatial resolution will still be limited by the need of obtaining reasonable large swath widths, as requested by most user applications. For Pan and MS sensors, this is obtained by accommodating long (or combining several shorter) linear arrays in the focal plane, possibly with the TDI technology. However, we remind that this strategy is not viable for current HS detectors and seems to be hardly feasible also in a future.

The limitation imposed by the transmission rate is easily understandable but its implications are somewhat hidden. A greater spatial and/or spectral resolution implies a greater volume of data to be stored onboard and transmitted, with a direct consequence on power and memory requirements. Data compression techniques can partially alleviate but cannot solve the problem. The huge volume of data an HS could produce appears as a hard limit for spatial resolution.

All these considerations suggest that image pansharpening, or fusion in general, will still remain a useful tool for HS data, the same as for MS data. *Qualitative* photo analysis of very high-resolution (VHR) images acquired on usually unavailable spectral bands appears promising and could be used in the same way pansharpened images are commonly and often unawarely utilized by users, as for example in Google Earth.

Concerning such *quantitive* applications as imaging spectroscopy, accurate spatial classification and detection of material components, only preliminary studies have been carried out because of limitations of the data that are currently available. In the next future, it is envisaged that the availability of HS datasets equipped with consistent ground truth and auxiliary data will allow a quantitative evaluation of HS pansharpening techniques to be achieved. In particular, the evaluation will primarily focus on adequacy of spatial/geometric detail injection models and validation of results on real applications.

The question now is: *how can HS data fusion be accomplished?*

There are several approaches that can be used for HS resolution enhancement, most of which have heritage in MS pansharpening. HS pansharpening, however, is more complex than that of MS data, for three main reasons:

- the spectral range of the Pan image, visible or visible and near infrared (V-NIR) does not match the whole wavelength acquisition range of the HS bands (V-NIR + SWIR);

- the spatial scale ratio between HS and Pan may be greater than four (typical case of pansharpening) and may not be a power of two, as in the case of Hyperion data and the Pan channel of ALI; thus direct application of wavelet-based (DWT and UDWT) fusion methods is not immediate;

- the data format of HS data may be crucial; unlike MS pansharpening, in which spectral radiance and reflectance as start format may be equivalent in terms of performances of fusion product, for HS pansharpening the presence of absorption bands and spectrally selective path radiance offsets, together with the decaying shape of solar irradiance, make fusion to be expedited if a reflectance data product is preliminarily available.

A further crucial point is the spectral fusion model required to preserve the spectral information of the data with lower-spatial resolution, or even to enhance it through the spectral unmixing of the coarse-resolution HS pixels, based on information extracted from the fine-resolution Pan data. Spatial details that are not available for HS bands have to be inferred through the model, starting from the high-spatial frequency components of Pan. The fusion model should be as simple as possible, in order to limit the computational complexity, and the model parameters should be spatially invariant, band dependent, and should be easily, yet accurately, estimated from the available dataset.

Whenever the V-NIR and SWIR datasets have different spatial resolutions (technological reasons impose that the resolution of the SWIR image cannot be greater than that of the V-NIR), it is possible to preliminarily synthesize spatially enhanced SWIR bands at the scale of the V-NIR image by exploiting the V-NIR band that is statistically most similar to each SWIR band [97]. Then the whole datacube, that is, original V-NIR and V-NIR-sharpened SWIR, both at the scale of V-NIR, is pansharpened by means of the Pan image. The concept of *bootstrap* fusion, whenever feasible, is undoubtedly advantageous over single-step fusion. The possibility of fusing V-NIR and SWIR data before fusion with the Pan image is performed has been recently investigated [228]. To overcome the problem that V-NIR and SWIR bands are spectrally disjoint, multiple V-NIR bands, each acting as an *enhancing Pan* for the SWIR band that is statistically most similar are exploited. In other words, each SWIR band will be preliminary fused with a *Pan* that is nothing else than the V-NIR band statistically most similar to it. Ultimately, fusion of this *intermediate* result with the original high-resolution Pan image is considered.

8.3   Literature Review

Thus far, most of developments of spatial resolution enhancement of HS data were rough extensions of MS pansharpening. This is the main reason that the literature on this topic is scarce and usually limited to conference proceedings, where applications of existing pansharpening methods to HS data are reported, rather than in journal articles, where it is expected that novel theoretical fusion approaches are experimented on HS data. Also the lack of suitable datasets (HS + Pan) further prevented from stimulating researchers' ideas.

Among the methods that can be easily extended to HS pansharpening, variations of component substitution (CS) and multiresolution analysis (MRA) methods are encountered [171]. Instead, other methods, have been specifically developed for HS data. They are based on unmixing concepts [129, 96], Bayesian estimation [130, 276], statistical optimization, for example, MMSE estimation, [60] and spatial distortion optimization [149, 150, 111]. Other methods use a sensor simulation strategy based on spectral response functions of the sensor to maintain the spectral characteristics of HS data [152]. Ultimately, methods that do not exploit a higher-spatial resolution dataset, like super-resolution mapping [187] and vector bilateral filtering [208] are beyond the scope of this book.

In the following, *band dependent spatial detail* (BDSD) MMSE fusion [60] and HS fusion with a constrained spectral angle (CSA) [111] are reported as examples of methods that try to overcome the spectral mismatch of the datasets and are thus able to produce fused HS data of adequate subjective and objective quality.

8.3.1   BDSD-MMSE Fusion

In HS pansharpening, a direct, unconditioned injection of spatial details extracted from Pan gives unsatisfactory results. A possible solution is to design an injection model providing the optimum strategy for spatial enhancement

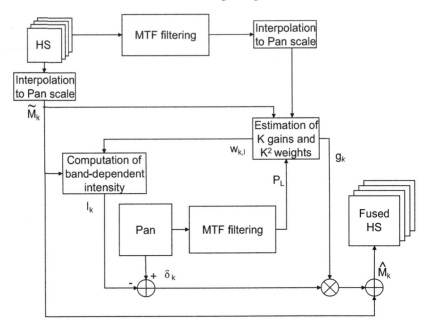

FIGURE 8.1: Flowchart of BDSD-MMSE fusion of HS data.

by simulating fusion at degraded resolution and minimizing the MSE between the original HS data and the fused result.

The MMSE method originally proposed in [60] for pansharpening of HS images has been successfully extended to the spatial enhancement of MS images [119]. An optimal detail image extracted from the Pan image is calculated for each HS band by evaluating the *band-dependent* generalized intensities.

The algorithm is optimal in the MMSE sense, since it solves a linear problem for *jointly* estimating the K^2 weights, $w_{k,l}$, defining the generalized intensities, $I_k = \sum_{l=1}^{K} w_{k,l}\tilde{M}_l$, and the K gains, g_k, that rule the injection of the spatial details; K denotes the number of HS bands and the indices k and l vary from 1 to K. The flowchart of the algorithm is shown in Figure 8.1. The mathematical model is the following:

$$\hat{\mathbf{M}}_k = \tilde{\mathbf{M}}_k + g_k \left(\mathbf{P} - \sum_{l=1}^{K} w_{k,l}\tilde{\mathbf{M}}_l \right) \qquad k = 1, \dots, K, \qquad (8.1)$$

where $\hat{\mathbf{M}}_k$ and $\tilde{\mathbf{M}}_k$, are the kth fused and original (interpolated to the scale of Pan) HS bands, respectively, and $\mathbf{P}$ is Pan. All images are column-wise lexicographically ordered for matrix notation. Eq. (8.1) may be rewritten in compact form

$$\hat{\mathbf{M}}_k = \tilde{\mathbf{M}}_k + \mathbf{H}\,\gamma_k, \qquad (8.2)$$

where $\mathbf{H} = \left[\tilde{\mathbf{M}}_1, \tilde{\mathbf{M}}_2, \dots, \tilde{\mathbf{M}}_k, \mathbf{P}\right]$ is the *observation matrix* of the linear model, and $\boldsymbol{\gamma}_k = [\gamma_{k,1}, \gamma_{k,2}, \ldots, \gamma_{k,K+1}]^T$ redefines the set of $K \times (K + 1)$ parameters to be estimated $(k = 1, \ldots, K)$:

$$\gamma_{k,l} = \begin{cases} -g_k \, w_{k,l} & l = 1, \ldots, K \\ g_k & l = K + 1. \end{cases} \tag{8.3}$$

The $K \times (K + 1)$ scalar parameters in $\boldsymbol{\gamma}_k$ are jointly optimally estimated by applying (8.1) at degraded resolution, that is, by spatially enhancing interpolated versions of the lowpass-filtered original MS bands, namely $\tilde{\mathbf{M}}_{k_L}$, by means of the lowpass-filtered Pan image, $\mathbf{P}_L$, to obtain an estimate, $\hat{\mathbf{M}}_{k_L}$, of the interpolated versions, $\tilde{\mathbf{M}}_k$, of the the original HS bands $\mathbf{M}_k$.

$$\hat{\mathbf{M}}_{k_L} = \tilde{\mathbf{M}}_{k_L} + g_k \left(\mathbf{P}_L - \sum_{l=1}^{K} w_{k,l} \tilde{\mathbf{M}}_{l_L} \right) \quad k = 1, \ldots, K, \tag{8.4}$$

Specific MTF-shaped Gaussian filters are considered for each MS band and for the Pan image [119]. The MMSE solution of (8.4) is given by the following *minimum variance unbiased estimator* (MVUE):

$$\boldsymbol{\gamma}_k = \left(\mathbf{H}_d^T \mathbf{H}_d \right)^{-1} \mathbf{H}_d^T \left(\tilde{\mathbf{M}}_k - \tilde{\mathbf{M}}_{k_L} \right), \tag{8.5}$$

in which $\mathbf{H}_d = \left[\tilde{\mathbf{M}}_{1_L}, \ldots, \tilde{\mathbf{M}}_{K_L}, \mathbf{P}_L \right]$ indicates the observation matrix at degraded resolution, but same spatial scale, constituted by the interpolated versions of the MTF-filtered original HS bands and the MTF-filtered Pan image. To save computation, the MMSE estimation at degraded resolution may be carried out at the scale of HS, instead of Pan. With reference to Figure 8.1, this is achieved by decimating $\mathbf{P}_L$, avoiding interpolation of $\mathbf{M}_{k_L}$ and using $\mathbf{M}_k$ instead of $\tilde{\mathbf{M}}_k$ as a reference in the estimation block of g_k and $w_{k,l}$.

In substance, the BDSD-MMSE algorithm is based on an unconstrained global MSE optimization carried out at a degraded spatial resolution, in which the original HS data serve as reference. The underlying assumption is that the optimal spectral weights and injection gains found at degraded resolutions are still optimal when fusion is performed at full resolution. A recent upgrade of the BDSD method consists of performing a space-varying nonlocal optimization of its parameters [109].

8.3.2 Fusion with a Constrained Spectral Angle

A desirable feature of an HS pansharpening method is to preserve the spectral information of the original HS data, simultaneously avoiding spatial over/underenhancement. This approach leads to the CSA method [111]. As shown in Chapter 7, the injection model that yields a constrained angle can be achieved either by CS or by MRA fusion methods. The latter are preferable in the context of HS pansharpening, because they avoid problems with spectral mismatches occurring with CS methods.

Let us start from the HPM fusion model (7.10) and introduce a multiplicative factor β aimed at magnifying or damping the spectral vector that constitutes the spatial enhancement:

$$\hat{\mathbf{M}}_k = \tilde{\mathbf{M}}_k + \beta \cdot \frac{\tilde{\mathbf{M}}_k}{\mathbf{P}_L}\left(\mathbf{P} - \mathbf{P}_L\right), \ \forall k = 1, \ldots, K, \qquad (8.6)$$

in which the coefficient β is chosen in order to minimize the *spatial* distortion D_S, for example, measured according to the QNR protocol (Eq. (3.15)). In Section 8.4, it will be shown that the optimal value of β may be accurately predicted from measurements of D_S.

The CSA method features the further advantage that, if the original HS data have been preliminarily corrected for atmospheric transmittance and path radiance as well as for solar irradiance, no change in the spectral angle of the fused pixel vector means that the spectral reflectance of the original HS dataset is thoroughly preserved. Trivially, this happens if the HS data are simply interpolated without addition of spatial details. In this case, however, fusion is ineffective, because the spatial enhancement is missing. This *void* fusion method will be referred as EXP in Section 8.4 and represents the reference for spectral quality in full-scale fusion. Unless a fusion method is capable of spectrally *unmixing* the coarse resolution HS pixels by means of the fine-resolution Pan data, the void fusion EXP yields the lowest angular error, or equivalently, the lowest spectral distortion.

8.4   Simulation Results

The hyperspectral pansharpening methods described in the previous sections have been assessed by taking image data collected on June 23, 2002 by the Hyperion spaceborne HS scanner on the area of Palo Alto, California. The 220 HS bands span the wavelength interval 0.4 μm to 2.5 μm, with a spatial resolution of 30 m. The Pan image has been acquired by the ALI scanner and approximately covers a short subinterval of HS (0.45 μm to 0.69 μm) with a spatial resolution of 10 m. The original Pan image is of size 540 × 1800 and the original HS image of size 180 × 600.

Quantitative performances have been computed by considering some score indices that require reference originals:

- BIAS between reference and fused HS images (3.1);

- Correlation coefficient (CC) to measure geometric distortion (3.5);

- Root mean square error (RMSE) as radiometric distortion index (3.3);

- Spectral-angle mapper (SAM), as a spectral distortion index (3.8);

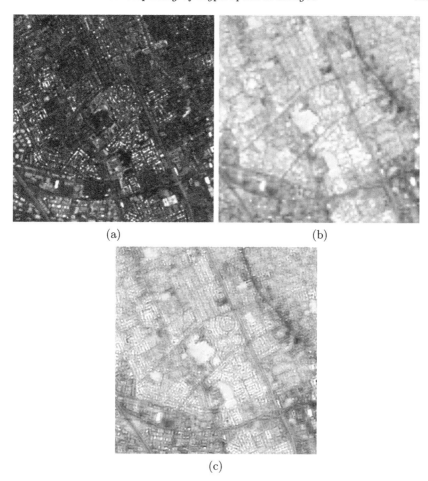

(a) (b)

(c)

FIGURE 8.2: Examples of full-scale spatial enhancement of fusion algorithms displayed as 300×300 false-color compositions: (Blue = band 17 (518.39 nm), Green = band 47 (823.65 nm) and Red = band 149 (1638.81 nm)). (a): ALI Pan Image; (b): resampled HS image; (c): BDSD-MMSE.

- ERGAS (3.10), as a cumulative normalized distortion index;

- $Q2^n$ [116], a generalization of Q4 (3.12) to more than four bands.

Fusion simulations have been performed both at the full scale of Pan (10 m) and at a reduced resolution (30 m) to check the synthesis property of Wald's protocol (see Section 3.2.2.1).

8.4.1 BDSD-MMSE Algorithm

In this simulation, BDSD-MMSE is compared with the following algorithms:

- Generalized intensity-hue-saturation method (GIHS) [252];

- Gram-Schmidt spectral sharpening method (GS) [157];

- UNB pansharp [275, 37], as implemented in PCI Geomatica software;

- Highpass filtering (HPF) with a 5×5 box filter suitable for $r = 3$ [66];

- Highpass modulation (HPM) with a 5×5 box filter [226, 174].

The visual appearance of a BDSD-MMSE fusion product at full scale can be checked in Figure 8.2.

Figure 8.3 shows reference and fused images at degraded resolutions in a false-color combination of bands.

Table 8.1 reports three scalar indices (average on bands) and two vector indices calculated for the Hyperion dataset at degraded resolutions.

TABLE 8.1: Scalar and vector distortion indices of Hyperion/ALI fusion computed at degraded resolutions.

	BIAS	CC	RMSE	ERGAS	SAM
REF	0	1	0	0	0
BDSD-MMSE	0.003	0.991	126.95	2.873	1.999
GIHS	-2.501	0.685	663.66	18.052	8.242
GS	-0.182	0.750	636.61	14.352	9.591
UNBPS	-0.133	0.976	240.13	8.421	6.893
HPF	-0.686	0.966	198.64	6.174	2.302
HPM	-0.128	0.985	151.25	3.427	1.938
EXP	-0.063	0.986	147.19	3.167	1.938

8.4.2 CSA Algorithm

The second simulation compares the CSA method with the BDSD-MMSE and other algorithms, some of them already considered in the previous simulation, and is especially devoted to finding the optimal value of the β parameter. First, one notes that the trend with the β parameter is almost identical for ERGAS, calculated from fused images and reference originals, and for the spatial/radiometric distortion index D_S of the QNR protocol (3.15). The latter is calculated without the aid of reference originals. From the plots in Figure 8.4, it stands out that the ERGAS minimizing value $\beta = 0.35$ can be accurately predicted as the minimum value in the plot of D_S versus β.

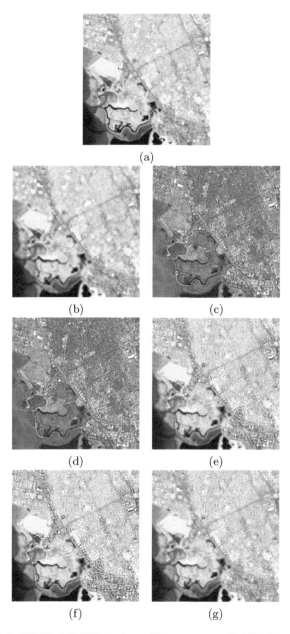

FIGURE 8.3: BDSD-MMSE fusion of Hyperion and ALI data at degraded resolutions (180 × 180 details at 30 m; false-color display, same as in Figure 8.2): (a): 30 m reference HS; (b): BDSD-MMSE; (c): GIHS; (d): GS; (e): UNBPS; (f): HPF; (g): HPM.

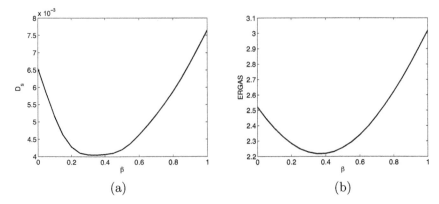

(a)   (b)

FIGURE 8.4: Distortions calculated from original and fused Hyperion HS image at 30 m varying with the β parameter. (a): QNR-D_S of fused 30 m HS, synthesized from 90 m HS and 30 m Pan; (b): ERGAS between fused 30 m HS, synthesized from 90 m HS and 30 m Pan, and original 30 m HS.

The CSA fusion method in all its versions ($\beta = 0$, EXP; $\beta = 1$, HPM; $\beta = 0.35$, CSA with spatial distortion optimization) has been compared with:

- BDSD-MMSE;

- GIHS;

- GS;

- UNB pansharp.

Figure 8.5 reports fused images for CSA and some compared algorithms, namely GIHS and GS (ENVI© implementation). The performances of CSA concern the algorithm running with the optimal value of $\beta = 0.35$ and with the boundary values $\beta = 1$, corresponding to HPM, and $\beta = 0$, corresponding to plain interpolation (EXP).

Table 8.2 reports SAM, ERGAS, and $Q2^n$ indices, which are calculated at a reduced resolution, for CSA and other methods that are jointly compared. Following the procedure adopted for the Q4 index, the $Q2^n$ index is calculated on 32×32 blocks and then averaged over all the blocks to yield a unique global value.

8.4.3   Discussion

In the first simulation, Table 8.1 shows that, for all five indices, the BDSD-MMSE method attains scores better than those of the other algorithms, followed by HPM, which retains the spectral quality (SAM) of the interpolated product (EXP) at the price of a slightly poorer radiometric quality (ERGAS). It is not surprising that BDSD-MMSE attains the best MMSE result, even

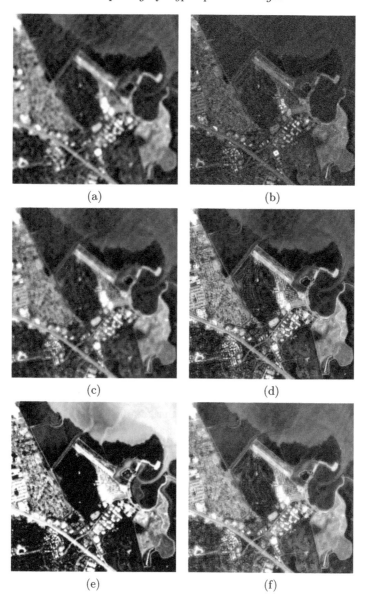

FIGURE 8.5: Details of original and fused Hyperion HS images (300 ×300 at 10 m scale). True-color display with Blue = band 11 (457 nm), Green = band 20 (549 nm), Red = band 30 (651 nm): (a): 30 m original Hyperion interpolated at 10 m (EXP in Table 8.1, corresponding to CSA with $\beta = 0$); (b): 10 m ALI Pan image; (c): CSA with optimal $\beta = 0.35$; (d): CSA with $\beta = 1$; (e): GIHS; (f): GS.

TABLE 8.2: Vector indices computed at degraded resolutions for a Hyperion/ALI fusion performance evaluation.

	SAM	ERGAS	$Q2^n$
REF	$0°$	0	1
BDSD-MMSE	$1.956°$	2.106	0.938
GIHS	$7.921°$	15.278	0.584
GS	$9.269°$	11.849	0.619
UNBPS	$6.572°$	7.127	0.862
CSA (w/ $\beta = 0.35$)	$1.617°$	2.217	0.932
HPM (CSA w/ $\beta = 1$)	$1.617°$	3.020	0.897
EXP (CSA w/ $\beta = 0$)	$1.617°$	2.521	0.919

better than EXP, at the price of a slightly poorer spectral performance. Nevertheless, it is noteworthy that the MMSE optimization also leads to benefits in the spectral quality. Both HPF and HPM rely on the same undecimated MRA, but the latter yields far lower distortions than the former, thanks to the angle-preserving injection model. This fact is relevant because it points out the difficulty of performing an adequate fusion for HS data. Only properly designed algorithms may obtain improvements, either spectral or radiometric, over the plain interpolation (EXP). A puzzling consideration arises from that: plain interpolation yields both spectral and radiometric fidelity to the reference 30 m HS data practically unsurpassed (BDSD-MMSE is the sole exception) by any other method, at least among those compared. Thus, it seems that spatial sharpening is accompanied by a loss of quality of the original HS data. CSA with optimized β represents the sole exception.

In Figure 8.3, one can note that GIHS (Figure 8.3(c)) and GS (Figure 8.3(d)) exhibit marked spectral distortions with respect to the reference (Figure 8.3(a)). This is not surprising, given their abnormally high error values in Table 8.1. UNB pansharp (Figure 8.3(e)), which is a kind of *smarter* GS, as well as HPF (Figure 8.3(f)) is affected by significant overenhancement. HPM (Figure 8.3(g)) is spectrally accurate but slightly too sharp with respect to the reference. BDSD-MMSE (Figure 8.3(b)) yield the most accurate results, both geometrically and spectrally. The visual inspection at full scale in Figure 8.2 highlights that all the spectral signatures of the original HS data are carefully incorporated in the sharpened bands. The BDSD-MMSE fusion product is geometrically rich and detailed as the high-resolution ALI image.

Unlike the first one, in the second simulation, SWIR absorption bands, which are responsible for the somewhat poor performances of all methods, have been removed. Table 8.2 provides evidence that the CSA method featuring the optimal value of $\beta = 0.35$ shows the best SAM values, together with EXP and HPM. However, HPM exhibits a larger radiometric distortion (ERGAS about 30% higher). EXP lies in the middle. It is noteworthy that the sole difference between CSA, HPM, and EXP is the value of β. Not surprisingly the BDSD-MMSE attains ERGAS 5% lower than that of CSA, but SAM more than 20% higher. Furthermore, for a 200-band image, the BDSD-MMSE method requires LS optimization of $200 \times (200 + 1) = 40200$ parameters at degraded resolution. The other three methods, namely GIHS, GS and UNBPS, though implemented in commercial software packages and hence widely used for photoanalysis applications, seem to be unsuitable for HS pansharpening, because of the very large radiometric and especially spectral errors that may be unacceptable for HS analysis. The visual results at full scale in Figure 8.5 are substantially in trend with the numerical values of indices.

Ultimately, the leftmost column of Table 8.2 reports $Q2^n$ [116], the extension of Q4, which seems to possess the favorable properties of Q4, in the sense of trading off the spectral (SAM) and geometric/radiometric distortions (ERGAS). According to $Q2^n$, BDSD-MMSE achieves the best rank ($Q2^n = 0.938$) closely followed by CSA with optimal β ($Q2^n = 0.932$). EXP is third with $Q2^n = 0.919$; HPM fourth with $Q2^n = 0.897$, because of its radiometric distortion. All the other methods (UNBPS, GS, and GIHS) are increasingly poorer, because of their abnormally high distortions. An immediate consideration is that such methods are based on CS, whose straightforward application to HS data may be questionable. The CSA methods compared, with the exception of the limit case EXP, exploit MRA instead of CS. BDSD-MMSE is an unconstrained LS minimization of a mathematical injection model, whose performances depend on the fundamental assumptions that the parametric estimation of the model may be carried out at degraded resolutions, which, in turn, depends upon on the landscape.

8.5   Concluding Remarks

With the present and upcoming availability of simultaneous acquisitions of an HS image and a higher resolution Pan image of the same scene, the problem of extending fusion methods for spatial resolution enhancement to HS datasets is becoming of great interest.

The peculiarity of high-spectral resolution imagery suggests distinguishing between *qualitative* applications of spatially enhanced HS images, such as photo analysis, and *quantitative* applications, such as imaging spectroscopy. In fact, different goals require different approaches to HS pansharpening: statistical optimization methods, such as MMSE estimation, provide high-quality sharpened images with high values of score indices, but may not exactly preserve the original spectral information of the low-resolution HS data, for which specific constraints need to be incorporated in the fusion algorithm, as in CSA schemes. On the other hand, most of the classical methods designed for MS pansharpening, like GIHS and Gram-Schmidt spectral sharpening, produce unsatisfactory results when applied to HS images, because they introduce considerable radiometric and spectral errors that may impair both visual and quantitative analyses.

Two-step, or *bootstrap*, fusion could represent a viable approach to HS pansharpening whenever the instruments in the V-NIR and SWIR wavelengths produce images having different spatial resolution from each other. When V-NIR data have greater resolution than SWIR data, one or more V-NIR bands can act as a sharpening image, the same as the Pan image, for each of the SWIR bands. The intermediate fusion products (SWIR sharpened by V-NIR) together with the original V-NIR bands may be further sharpened if a Pan image is available. The first step of this approach is of interest for airborne systems, whenever the V-NIR and SWIR instruments have different spatial resolutions and a Pan image is unavailable.

Ultimately, the huge volume of HS data makes the computational complexity of pansharpening methods based on sophisticated radiative transfer models and/or constrained mathematical optimization, like most of those reviewed in Chapter 11, definitely not affordable for HS fusion, at least thus far.

Chapter 9

Effects of Aliasing and Misalignments on Pansharpening

9.1 Introduction

As explained in Chapter 7, the majority of pansharpening methods belongs to two main classes that uniquely differ in the way the spatial details are extracted from the Pan image. By summarizing, these two classes can be described as:

- Techniques that employ linear space invariant digital filtering of the Pan image to extract the spatial details, i.e., the geometrical information, that will be added to the MS bands [216]; all methods employing MRA belong to this class.

- Techniques that yield the spatial details as pixel difference between the Pan image and a nonzero-mean component obtained from a spectral transformation of the MS bands, without any spatial filtering of the former. They are equivalent to substitution of such a component with the Pan image followed by reverse transformation to produce the sharpened MS bands [26]. CS-based methods, if they are not coupled with MRA, belong to this class.

A topic rarely investigated in literature is the sensitivity of pansharpening methods, in the presence of the two most common impairments occurring with fusion of VHR products, aliasing and spatial/temporal misalignments, which can differently affect the sharpened products depending on the class of algorithms used in the fusion procedure [15, 45, 10].

Aliasing is generally not present in the Pan image, which is acquired with a low value of MTF at the Nyquist frequency and digitally postprocessed on the ground for MTF restoration. Instead, aliasing is a common problem of all MS bands and originates from an insufficient sampling step size, or equivalently by a too high value of the MTF at Nyquist. As a consequence, annoying jagged edges, especially noticeable on sharp straight oblique contours, may appear in the original MS images.

On the other hand, spatial misalignments can be caused by registration inaccuracies of geocoded products. Misregistration or improper interpolation of the MS bands generate undesired shifts between the resampled MS dataset and the Pan image, as shown in Chapter 4. Temporal misalignments may occur if the fusion procedure uses MS and Pan of the same scene taken at different times and possibly by different platforms.

In this chapter, the effects of aliasing and spatial/temporal misalignments on pansharpened imagery are investigated, both theoretically and experimentally, for both CS- and MRA-based fusion methods. Starting from the mathematical formulation of the two main approaches, it will be proven that under general and likely assumptions, CS methods are almost insensitive to aliasing and less sensitive than MRA methods to spatial misalignments of a moderate extent. However, within MRA-based methods, GLP is also capable of compensating the aliasing of MS, unlike ATW and analogously to CS methods, thanks to the decimation and interpolation steps present in its flowchart [12]. Eventually, both quantitative and qualitative results show that multitemporal misalignments are better compensated by MRA rather than by CS methods, conversely from spatial misalignments.

9.2   Mathematical Formulation

With reference to Figures 7.1 and 7.4, let us denote, as usual, the original low-resolution MS dataset comprising N bands as $\tilde{\mathbf{M}}_k$, $k = 1, \ldots, N$, once expanded to the scale of the Pan image, and as $\hat{\mathbf{M}}_k$, $k = 1, \ldots, N$ the high-resolution dataset of the fused MS bands. The Pan image and its lowpass version (possibly decimated and successively interpolated) are denoted as $\mathbf{P}$ and $\mathbf{P}_L$, respectively. The gain matrix $\mathbf{G}_k$ modulates the highpass spatial detail to be injected into each MS band according to the adopted model, either global or local, while the coefficient set $\{w_k\}$ is adopted by CS-based methods to synthesize the smooth intensity $\mathbf{I}$. A switch element enables CS- or MRA-based frameworks.

Let us consider the general equation for CS- and MRA-based approaches:

$$\hat{\mathbf{M}}_k = \tilde{\mathbf{M}}_k + \mathbf{G}_k \cdot \delta, \quad k = 1, \ldots, N \tag{9.1}$$

where δ represent the details of the Pan image to be injected in the interpolated MS bands and $\mathbf{G}_k$ the image-dependent matrix that weighs the Pan details, through a point-to-point multiplication. For ease of notation, the gain matrix $\mathbf{G}_k$ is assumed to be constant throughout the kth band. Starting from (9.1), the schemes of Figures 7.1 and 7.4 can be translated into simple equations describing how the two families work. These equations are briefly recalled here as starting points for the subsequent developments.

9.2.1   CS-Based Methods

CS-based methods are described by:

$$\hat{\mathbf{M}}_k = \tilde{\mathbf{M}}_k + \mathbf{G}_k \cdot (\mathbf{P} - \mathbf{I}), \quad k = 1, \ldots, N \tag{9.2}$$

where $\delta = \mathbf{P} - \mathbf{I}$ and $\mathbf{I}$ is given by a linear combination of the MS dataset, whose coefficients depend on the chosen transformation, that is:

$$\mathbf{I} = \sum_{k=1}^{N} w_k \cdot \tilde{\mathbf{M}}_k, \quad k = 1, \ldots, N \tag{9.3}$$

where the weights $\{w_k\}$ change for different pansharpening methods.

9.2.2   MRA-Based Methods

The proposed model for MRA-based methods can be written as:

$$\hat{\mathbf{M}}_k = \tilde{\mathbf{M}}_k + \mathbf{G}_k \cdot (\mathbf{P} - \mathbf{P}_L), \quad k = 1, \ldots, N, \tag{9.4}$$

with $\delta = \mathbf{P} - \mathbf{P}_L$. For the ATW method, the lowpass-filtered Pan version $\mathbf{P}_L$ is given by $\mathbf{P}_L = \mathbf{P} \otimes \mathbf{h}$, where the lowpass filter $\mathbf{h}$ can be obtained by means of the MTF of each band. In this case, (9.4) modifies into:

$$\hat{\mathbf{M}}_k = \tilde{\mathbf{M}}_k + \mathbf{G}_k \cdot (\mathbf{P} - \mathbf{P} \otimes \mathbf{h}) = \tilde{\mathbf{M}}_k^* + \mathbf{G}_k \cdot \mathbf{P} \otimes \mathbf{g}, \quad k = 1, \ldots, N, \quad (9.5)$$

in which $\mathbf{g}$ is the equivalent highpass filter of ATW.

In the case of GLP fusion, if q is the scale ratio between the Pan and the MS images, $\mathbf{P}_L$ is obtained by filtering, degrading, and successively interpolating the $\mathbf{P}$ image, so that (9.4) becomes:

$$\hat{\mathbf{M}}_k = \tilde{\mathbf{M}}_k + \mathbf{G}_k \cdot (\mathbf{P} - expand_q((\mathbf{P} \otimes \mathbf{h}) \downarrow q)), \quad k = 1, \ldots, N. \quad (9.6)$$

In this analysis, the lowpass filter $\mathbf{h}$ is assumed to be identical to the MTF of the kth spectral channel, as in (9.5).

9.3 Sensitivity to Aliasing

Let us consider $\tilde{\mathbf{M}}_k$, that is, the kth band of the MS image interpolated at the spatial scale of the $\mathbf{P}$ image. The term $\tilde{\mathbf{M}}_k$ is supposed to be generated by the superposition of an interpolated aliasing free image, $\tilde{\mathbf{M}}_k^*$, and of an interpolated aliasing pattern, $\tilde{\mathbf{A}}_k$. The general expression (9.1) of the fused kth band, that is, $\hat{\mathbf{M}}_k$, becomes:

$$\hat{\mathbf{M}}_k = \tilde{\mathbf{M}}_k + \mathbf{G}_k \cdot \delta = \tilde{\mathbf{M}}_k^* + \tilde{\mathbf{A}}_k + \mathbf{G}_k \cdot \delta \quad (9.7)$$

9.3.1 CS-Based Methods

For CS-based methods, Eq. (9.2) is changed into:

$$\hat{\mathbf{M}}_k = \tilde{\mathbf{M}}_k + \mathbf{G}_k \cdot (\mathbf{P} - \mathbf{I}) = \tilde{\mathbf{M}}_k^* + \tilde{\mathbf{A}}_k + \mathbf{G}_k \cdot (\mathbf{P} - \mathbf{I}) \quad (9.8)$$

where the generalized intensity component $\mathbf{I}$ is given by a linear regression of the spectral bands, with weights $\{w_k\}$, by following (9.3). In the presence of aliasing, we have:

$$\mathbf{I} = \sum_k w_k \cdot \tilde{\mathbf{M}}_k = \sum_k w_k \cdot \tilde{\mathbf{M}}_k^* + \sum_k w_k \cdot \tilde{\mathbf{A}}_k \quad (9.9)$$

Hence, Eq. (9.8) can be rewritten as:

$$\hat{\mathbf{M}}_k = \tilde{\mathbf{M}}_k^* + \tilde{\mathbf{A}}_k + \mathbf{G}_k \cdot (\mathbf{P} - \sum_i w_i \cdot \tilde{\mathbf{M}}_i^* - \sum_i w_i \cdot \tilde{\mathbf{A}}_i). \quad (9.10)$$

In the assumption that $\mathbf{G}_k \cdot \sum_i w_i \tilde{\mathbf{A}}_i = \tilde{\mathbf{A}}_k$, which is reasonable since interpolated aliasing patterns of spectrally correlated bands are similar, and $\mathbf{G}_k \cdot \sum_i w_i \approx 1$, which is exactly true for GIHS (Tu *et al.* [251]), as shown in Aiazzi *et al.* [25], (9.10) becomes:

$$\hat{\mathbf{M}}_k = \tilde{\mathbf{M}}_k^* + \mathbf{G}_k \cdot \left(\mathbf{P} - \sum_i w_i \cdot \tilde{\mathbf{M}}_i^*\right) = \hat{\mathbf{M}}_k^*. \tag{9.11}$$

That is, the sharpened images produced from aliased MS bands (and aliasing free Pan) are identical to the products generated by the same algorithm starting from MS and Pan that are both aliasing free. This means that CS-based methods described by (9.2) are capable of exactly recovering aliasing impairments in the ideal case or, more realistically, to have a very low sensitivity to them.

9.3.2 MRA-Based Methods

In order to properly model the difference between ATW and GLP, let us reference Figures 7.6 and 7.7. For MRA-based methods, (9.4) becomes:

$$\hat{\mathbf{M}}_k = \tilde{\mathbf{M}}_k + \mathbf{G}_k \cdot (\mathbf{P} - \mathbf{P}_L) = \tilde{\mathbf{M}}_k^* + \tilde{\mathbf{A}}_k + \mathbf{G}_k \cdot (\mathbf{P} - \mathbf{P}_L) \tag{9.12}$$

where the term $\tilde{\mathbf{M}}_k$ is supposed, as in (9.8), to be generated by the superposition of an interpolated aliasing-free image, $\tilde{\mathbf{M}}_k^*$, and of an interpolated aliasing pattern, $\tilde{\mathbf{A}}_k$.

9.3.2.1 ATW-based fusion

For ATW-based fusion, the low-resolution Pan version $\mathbf{P}_L$ is obtained by lowpass filtering, that is, $\mathbf{P}_L = \mathbf{P} \otimes \mathbf{h}$, where $\mathbf{h}$ is the lowpass filter defining ATW, possibly equal to the MTF of the kth spectral channel. Consequently, (9.12) modifies into:

$$\hat{\mathbf{M}}_k = \tilde{\mathbf{M}}_k^* + \tilde{\mathbf{A}}_k + \mathbf{G}_k \cdot (\mathbf{P} - \mathbf{P} \otimes \mathbf{h}) = \tilde{\mathbf{M}}_k^* + \tilde{\mathbf{A}}_k + \mathbf{G}_k \cdot \mathbf{P} \otimes \mathbf{g} \tag{9.13}$$

in which $\mathbf{g}$ is the highpass filter of the ATW fusion. By considering (9.5), (9.13) can be rewritten as:

$$\hat{\mathbf{M}}_k = \tilde{\mathbf{M}}_k^* + \tilde{\mathbf{A}}_k + \mathbf{G}_k \cdot \mathbf{P} \otimes \mathbf{g} = \hat{\mathbf{M}}_k^* + \tilde{\mathbf{A}}_k \tag{9.14}$$

thus showing that the ATW fused image produced starting from aliased MS bands (and aliasing free Pan) contains the same aliasing pattern that appears in the interpolated MS image. Therefore, in this case, the aliasing pattern has not been compensated in the fusion process.

9.3.2.2 GLP-based fusion

Concerning GLP-based fusion, Eq. (9.12) becomes:

$$\hat{\mathbf{M}}_k = \tilde{\mathbf{M}}_k^* + \tilde{\mathbf{A}}_k + \mathbf{G}_k \cdot (\mathbf{P} - expand_q((\mathbf{P} \otimes \mathbf{h}) \downarrow q)). \tag{9.15}$$

If we assume that the lowpass filter $\mathbf{h}$ is identical to the MTF of the kth spectral channel, so that when its output is decimated by q, it would approximatively generate the same aliasing pattern of the one which has been produced by the MTF filter in the acquisition of the kth MS band, the following relation holds:

$$\mathbf{P} - expand_q((\mathbf{P} \otimes \mathbf{h}) \downarrow q) = \mathbf{P} - \mathbf{P}_L^* - \tilde{\mathbf{A}}_P, \qquad (9.16)$$

in which $\mathbf{P}_L^* \neq \mathbf{P}_L$, if the filter $\mathbf{h}$ exhibits a frequency response that is nonzero beyond one qth of the Nyquist frequency, as it generally happens with the MTF of the instrument. The term $\tilde{\mathbf{A}}_P$ represents the interpolated version of the aliasing pattern generated by decimation. The further assumption that $\tilde{\mathbf{A}}_k = \mathbf{G}_k \cdot \tilde{\mathbf{A}}_P$, which is likely because $\mathbf{G}_k$ rules the transformation of Pan details into MS details, yields:

$$\begin{aligned} \hat{\mathbf{M}}_k &= \tilde{\mathbf{M}}_k^* + \tilde{\mathbf{A}}_k + \mathbf{G}_k \cdot (\mathbf{P} - \mathbf{P}_L^* - \tilde{\mathbf{A}}_P) = \\ &= \tilde{\mathbf{M}}_k^* + \tilde{\mathbf{A}}_k + \mathbf{G}_k \cdot \mathbf{P} \otimes \mathbf{g}^* - \mathbf{G}_k \cdot \tilde{\mathbf{A}}_P = \tilde{\mathbf{M}}_k^* + \mathbf{G}_k \cdot \mathbf{P} \otimes \mathbf{g}^* \end{aligned} \qquad (9.17)$$

in which the frequency response of the equivalent highpass filter of GLP, $\mathbf{g}^*$, is identical to that of $\mathbf{g}$ (the highpass filter of ATW) except in the (positive) frequency interval between zero and one qth of the Nyquist frequency of Pan, where the former is identically zero, the latter is nonzero. If the previous approximations hold, by comparing with (9.14), it is apparent that the fusion products of the GLP method with MTF-matched reduction filter in the presence of aliased MS (and aliasing-free Pan) is identical to the products generated by the same algorithm starting from MS and Pan that are both aliasing free, as for GS-based methods. The difference is that the compensation is direct for GS-based methods, because it is made on the same patterns of aliasing. Conversely, the compensation is indirect for the MRA-GLP method, because it is made on the patterns of original MS bands and of the lowpass-filtered Pan image, whose aliasing patterns are only successively reported to the MS. In particular, a noticeable issue is represented by the phase difference caused by the slight delay between the acquisitions of the Pan image and the MS bands. In fact, the resulting phase shift can result into differences in the aliasing patterns, which cannot thus be fully compensated. This fact can be also space-dependent or scene-dependent, so that it is possible to remove the aliasing pattern in some areas and not in other ones. Consequently, the compensation of the GLP method is statistically less robust than that of the CS-based methods.

9.3.3   Results and Discussion

Fusion simulations will be devoted to demonstrate, both visually and graphically, the theoretical results which have been obtained in the previous sections. For this purpose, images acquired by VHR sensors, as IKONOS and QuickBird will be considered at the full scale. Quality and distortion indices will be used at the degraded scale for these sensors, so that quantitative plots can be shown to obtain an objective assessment. Finally, simulated Pléiades data allow objective measurements to be obtained also at the full scale thanks to the availability of a high-resolution MS reference image.

In the case of aliasing distortion, two distinct tests will be considered. The former one gives a visual comparison between GSA, a CS-based method, and MRA-ATW, in order to show the advantages of CS-based algorithms in rejecting aliasing. The latter comprises visual and quantitative results, which concern the previously cited methods together with an MRA-GLP fusion procedure. The objective is to show how GLP method is able to obtain performances comparable to CS-based methods in presence also of strong quantitative of aliasing patterns in the scene.

9.3.3.1   CS vs. MRA-ATW

The first set of simulations has been carried out on VHR data of the QuickBird sensor. In this test, a comparison is given between a CS-based method and an MRA-based one concerning ATW fusion. The first method is based on Gram-Schmidt spectral sharpening (GSA) [26], in particular it is a modified version of ENVI© GS [157] with improved spectral quality, thanks to a least squares (LS) calculation of the spectral weights w_k defining the generalized intensity $\mathbf{I}$. The second method is based on ATW [196, 200] with an MTF-matched Gaussian-like reduction filter [14]. The injection model, given by the matrix $\mathbf{G}_k$, is global in both cases and equal to the covariance between interpolated kth MS band and either $\mathbf{I}$ or the lowpass-filtered Pan ($\mathbf{P}_L$), divided by the variance of either $\mathbf{I}$ or $\mathbf{P}_L$ [25].

QuickBird data have been chosen for this set of simulations, since the amount of aliasing is related to the value of the bell-shaped MTF at Nyquist frequency, which is greater for QuickBird than for IKONOS [218, 75]. Figures 9.1(a)−(d) and 9.2(a)−(d) report two distinct tests. The former is related to a rural scene comprising fields and roads; the latter shows an urban landscape, essentially roofs. In both cases, the start-ups are aliasing-free Pan and aliased MS. Aliasing is due to insufficient sampling and is noticeable around sharp oblique contours of the scenes as annoying jagged patterns. Such patterns clearly appear in Figures 9.1 and 9.2 (b) interpolated at the spatial scale of Pan. The results of the two methods, GSA and ATW, are very different: the former rejects aliasing almost totally; the latter leaves aliasing patterns almost unchanged. Spectral quality is always good, thanks to the good properties of all MRA-based methods in general and of GSA in particular: no significant change in color hues is perceivable in all fusion products. However, under the perspective of aliasing rejection, this test shows that a CS method optimized for spectral quality [26] is preferable to an MRA-based method, yet optimized for spatial quality [14], in particular, the ATW method. Conversely, the following test will show that the GLP method can reach the same results of GSA in the presence of aliasing.

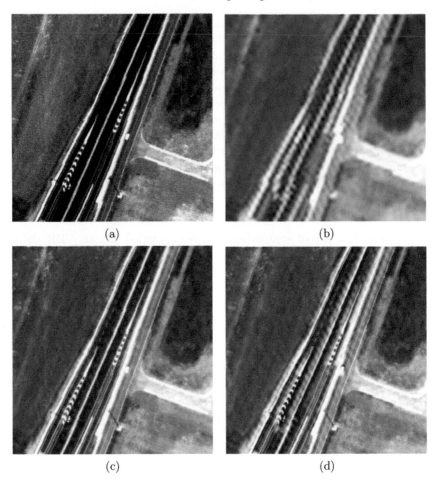

FIGURE 9.1: Details of original and fused QuickBird data (256×256 at 0.7 m scale): rural landscape. (a): panchromatic; (b): true-color display of 2.8 m MS interpolated at 0.7 m; (c): enhanced Gram-Schmidt; (d): "à-trous" wavelet with an MTF-matched filter and global injection model.

9.3.3.2 MRA-GLP vs. MRA-ATW

The results obtained in this simulation have been further analyzed by considering fusion simulations at degraded spatial scale and adding the MRA-GLP fusion method. In this test, GLP and ATW-based methods, both using the same MTF filters [14] and the same global injection gain [25], will be compared to show the advantages of GLP over ATW in the presence of aliasing of the MS bands. At the same time, the above two MRA-based methods will be compared also with GSA, that is, a CS-based method. The advantage of considering degraded images is twofold. On the one side, different amounts of

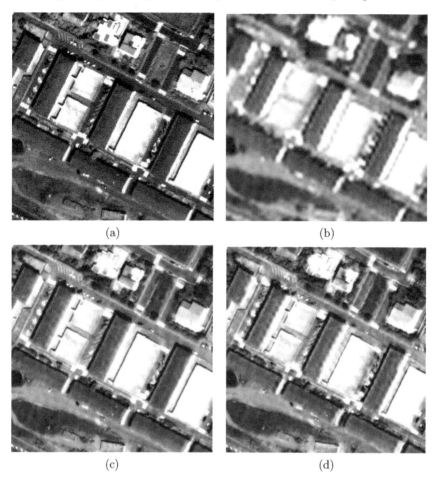

(a) (b)

(c) (d)

FIGURE 9.2: Details of original and fused QuickBird data (256×256 at 0.7 m scale): urban landscape. (a): panchromatic; (b): true-color display of 2.8 m MS interpolated at 0.7 m; (c): enhanced Gram-Schmidt; (d): "à-trous" wavelet with an MTF-matched filter and global injection model.

aliasing can be generated by tuning the passband of the lowpass filter preceding the downsampler. On the other side, fusion simulations carried out on degraded data allow reference originals to be used for quality evaluation according to Wald's protocol [261]. In this case, a true reference being available, quality/distortion indices can be used, namely: the average spectral angle mapper (SAM), measuring spectral distortion to originals of fusion products; the ERGAS index [216], measuring the average radiometric distortion; the Q4 index [33], jointly measuring the spectral and radiometric quality of fusion.

These simulations have been carried out by choosing IKONOS data, which are characterized by an amount of intrinsic aliasing lower than that of Quick-

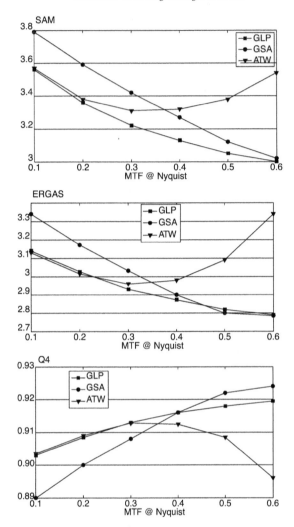

FIGURE 9.3: Quality/distortion indices measuring the sensitivity of the performances of GSA, GLP, and ATW methods to an increasing amount of aliasing, which is obtained by changing the amplitude at the Nyquist frequency of the Gaussian-like lowpass filter simulating the MTFs of the MS instrument.

bird data. Figure 9.3 reports SAM, ERGAS, and Q4 indices calculated between fusion products at 4 m, achieved from MS and Pan data spatially degraded by four along rows and columns, and original 4 m MS. Progressively increasing amounts of aliasing have been generated by applying Gaussian-shaped lowpass filters, with moduli of frequency responses at a cutoff Nyquist frequency increasing from 0.1 to 0.6 at steps of 0.1, before decimation is performed.

The MTFs of MS scanners generally stem from a trade-off between pass-bands and aliasing: the more the amount of signal in the passband, the more the amount of aliasing that occurs. Note that the original IKONOS data exhibit MTF values of spectral channels slightly lower than 0.3 [75]. Not surprisingly, all fusion methods benefit from aliasing up to a simulated MTF approximately equal to 0.3. After that value, GSA steadily follows the same trend leading to better and better fusion performances, as otherwise found in Baronti *et al.* [45]. Conversely, the two MRA-based methods, ATW and GLP, follow opposite trends leading to poorer and poorer performances of ATW and better and better performances of GLP, as the amount of aliasing increases. It is noteworthy that in Baronti *et al.* [45], it is proven that GSA, and analogous CS-based methods, remove aliasing, while ATW does not.

Eventually, simulated Pleiadés data have been used for the presentation of full-scale results. A Pan image at 0.7 m has been synthesized from the G and R channels of an airborne instrument. MS data at 2.8 m have been obtained by applying a laboratory model of the Pleiadés' MTF to the airborne data and decimating the outcome by four. Figure 9.4 shows 0.7 m synthetic Pan, 2.8 MS (true color) resampled to 0.7 m with noticeable aliasing impairments and 0.7 m true airborne MS for reference. The trends established in Figure 9.3 for the quantitative indices SAM, ERGAS, and Q4 is fully confirmed also for a visual assessment. Actually, the fusion products of GLP and GSA exhibit no perceivable aliasing. Conversely, ATW leaves most of aliasing patterns in the pansharpened image. This is especially noticeable on the street, where the spatial frequencies are distorted in the direction of the higher variability, as in Figure 9.1.

9.4   Sensitivity to Spatial Misalignments

Now let us consider the case of spatial misalignments between MS and Pan data. Let us denote with $(\Delta x, \Delta y)$ the spatial misalignment between the $\mathbf{P}$ image and the MS band interpolated at the scale of $\mathbf{P}$, that is, $\tilde{\mathbf{M}}_k$, with $\hat{\mathbf{M}}_k$ the fused kth band and with $\hat{\mathbf{M}}_k(\Delta x, \Delta y)$ the fused kth band in the presence of the misalignment $(\Delta x, \Delta y)$. Whether the misalignment is space-dependent and/or band-dependent has no impact on the validity of the following analysis. For a pansharpening algorithm showing a low sensitivity to a moderate spatial misalignment, the following relation should hold:

$$\hat{\mathbf{M}}_k(\Delta x, \Delta y) \approx \hat{\mathbf{M}}_k \qquad (9.18)$$

Starting again from the general expression for fusion, in the case of a spatial misalignment $(\Delta x, \Delta y)$, (9.1) becomes:

$$\hat{\mathbf{M}}_k(\Delta x, \Delta y) = \tilde{\mathbf{M}}_k(\Delta x, \Delta y) + \mathbf{G}_k \cdot \delta(\Delta x, \Delta y). \qquad (9.19)$$

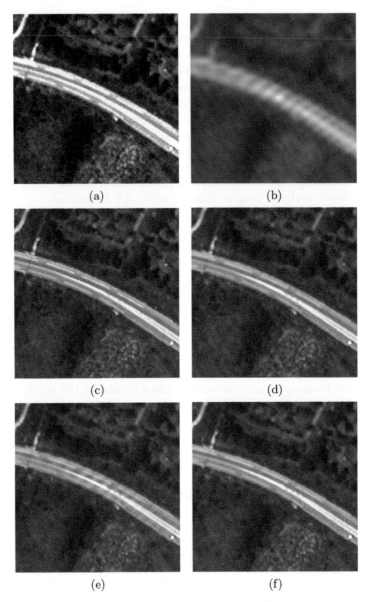

(a)   (b)

(c)   (d)

(e)   (f)

FIGURE 9.4: Details of simulated Pléiades data (256×256 at 0.7 m scale), original and fused. (a): synthetic Pan = (G+R)/2; (b): 2.8 m MS bicubically interpolated at 0.7 m; (c): original 0.7 m MS for reference; (d): generalized Laplacian pyramid (GLP) with an MTF-matched reduction filter and an almost-ideal 23-tap interpolator; (e): "à-trous" wavelet (ATW) transform with an MTF-matched filter; (f): enhanced Gram-Schmidt (GSA) [26].

If the misalignment $(\Delta x, \Delta y)$ is small, we can assume that $\tilde{\mathbf{M}}_k(\Delta x, \Delta y)$ can be approximated by a first-order Taylor expansion around the undistorted point $(\Delta x = 0, \Delta y = 0)$, so that we obtain:

$$\tilde{\mathbf{M}}_k(\Delta x, \Delta y) \approx \tilde{\mathbf{M}}_k + \mathbf{D}_x^{(k)}\Delta x + \mathbf{D}_y^{(k)}\Delta y, \tag{9.20}$$

in which the matrices $\mathbf{D}_x^{(k)}$ and $\mathbf{D}_y^{(k)}$ represent the derivatives of $\tilde{\mathbf{M}}_k$ in the x and y directions, along which the displacements $(\Delta x, \Delta y)$ occur. By considering Eq. (9.20), (9.19) can thus be rewritten as:

$$\hat{\mathbf{M}}_k(\Delta x, \Delta y) = \tilde{\mathbf{M}}_k + \mathbf{D}_x^{(k)}\Delta x + \mathbf{D}_y^{(k)}\Delta y + \mathbf{G}_k \cdot \delta(\Delta x, \Delta y) \tag{9.21}$$

9.4.1 MRA-Based Methods

For MRA-based fusion methods, the difference term $\delta = \mathbf{P} - \mathbf{P}_L$ is unaffected by spatial misalignments between the Pan image and MS bands, that is, $\delta(\Delta x, \Delta y) = \delta$. Therefore, (9.1) and (9.21) can be merged into the relation:

$$\hat{\mathbf{M}}_k(\Delta x, \Delta y) = \hat{\mathbf{M}}_k + \mathbf{D}_x^{(k)}\Delta x + \mathbf{D}_y^{(k)}\Delta y. \tag{9.22}$$

9.4.2 CS-Based Methods

For CS-based fusion methods, the difference term $\delta = \mathbf{P} - \mathbf{I}$ is affected by spatial misalignments, since $\mathbf{I}$ is a weighted average of the interpolated MS bands that are misaligned to the Pan image. If we denote with $\mathbf{I}(\Delta x, \Delta y)$ the misaligned intensity,

$$\delta(\Delta x, \Delta y) = \mathbf{P} - \mathbf{I}(\Delta x, \Delta y) \tag{9.23}$$

from which

$$\delta(\Delta x, \Delta y) \approx \mathbf{P} - \mathbf{I} - \mathbf{D}_x^{(I)}\Delta x - \mathbf{D}_y^{(I)}\Delta y \tag{9.24}$$

where the matrices $\mathbf{D}_x^{(I)}$ and $\mathbf{D}_y^{(I)}$ contain the discrete derivatives of $\mathbf{I}$ along x and y, respectively. Finally,

$$\delta(\Delta x, \Delta y) = \delta - \mathbf{D}_x^{(I)}\Delta x - \mathbf{D}_y^{(I)}\Delta y \tag{9.25}$$

By substituting (9.25) in (9.21) we obtain:

$$\begin{aligned}\hat{\mathbf{M}}_k(\Delta x, \Delta y) = \\ = \hat{\mathbf{M}}_k + (\mathbf{D}_x^{(k)} - \mathbf{G}_k \cdot \mathbf{D}_x^{(I)})\Delta x + (\mathbf{D}_y^{(k)} - \mathbf{G}_k \cdot \mathbf{D}_y^{(I)})\Delta y.\end{aligned} \tag{9.26}$$

The sensitivities of MRA- and CS-based pansharpening methods to spatial misalignments are resumed in (9.22) and (9.26), which will be discussed more in-depth. For small misalignments, the impairments are proportional to the displacement $(\Delta x, \Delta y)$, by means of some coefficients containing the derivatives of $\tilde{\mathbf{M}}_k$ and $\mathbf{I}$ along the x and y directions. Consequently, for homogeneous areas, in which the values of derivatives are small, the impairments are not significant. Conversely, for textured areas and contours, which are characterized by not negligible values of spatial derivatives, noticeable impairments may appear. However, this effect is much stronger for filter-based (MRA) techniques than for projection-based (CS) algorithms. In fact, by comparing (9.22) and (9.26), it is evident that in (9.26) the misalignments are compensated by the differences between the two terms including the derivatives of $\tilde{\mathbf{M}}_k$ and $\mathbf{I}$, respectively. In principle, if $\mathbf{D}_x^{(k)} = \mathbf{D}_x$ and $\mathbf{D}_y^{(k)} = \mathbf{D}_y$, $\forall k$, in the further assumption that $\mathbf{G}_k \cdot \sum_i w_i = 1$, the fused images with and without misalignment would be identical because the term in the second line of (9.26) would be identically zero. In practice, the derivatives are different by changing the band, though likely to have same sign and similar amplitudes, at least in the visible wavelengths. Therefore, the compensation of misalignment impairments in CS-based fusion methods may be less effective than that in the case of aliasing distortion. However, MRA-based algorithms have a higher sensitivity to spatial misalignments than CS-based methods.

9.4.3   Results and Discussion

Spatial misalignments are addressed by considering two different sources of displacements: the former evidences the effects of a misregistration in the acquisition of MS and Pan images; the latter is related to an unsuitable interpolation generating shifts which can affect the results of a pansharpening procedure.

9.4.3.1   Misregistration

In this test, an IKONOS dataset has been chosen for simulations. The reason is the low amount of aliasing of the IKONOS data, so that the possibility of masking the effects of the spatial misalignments can be avoided. The MS and Pan datasets have been misregistered along both x and y by 4 pixels at P scale, that is, one pixel at MS scale. The compared fusion methods are still GSA and ATW. Two distinct tests are shown in Figures 9.5 and 9.6. The former concerns an agricultural field with a bright circle, the latter shows a detail of a little town with many roofs. For each test, six images have been displayed: the Pan, the interpolated MS, GSA fusion of overlapped MS and Pan, GSA fusion of misregistered MS and Pan, ATW fusion of overlapped data and ATW fusion of misaligned data. The visual results are stunning. While GSA and ATW behave quite similarly in absence of misregistration, on misaligned data, the former produces an image with high resemblance to the ideal case, apart from colors, which are obviously shifted together with the MS original.

The geometry of spatial details is preserved to a large extent notwithstanding the 4-pel shift. Conversely, ATW exhibits a better preservation of colors, but can not tolerate the 4-pel shift, while in Figure 9.5(f) the circular target is split into its lowpass and edge components. In Figure 9.6(f) the geometry of the roof scene fades off.

TABLE 9.1: Geometrical (D_S) and spectral (SAM) distortions of fusion based on MRA (ATW) and on CS (GSA). IKONOS test image, 1024×1024. Interpolated MS and Pan exactly overlapped and misregistered (MIS) by 4 pels along x and y.

	ATW	MIS ATW	GSA	MIS GSA
D_S	0.0860	0.2869	0.0974	0.2017
SAM	0.4359	3.1315	0.4691	3.1562

In order to quantify the losses in quality, either spectral or spatial/geometric, of the two sample methods, GSA and ATW, in the presence of misregistration, separate measurements of SAM and of spatial distortion, D_S, according to the QNR protocol [32] have been performed between fused and interpolated MS data. The choice of distortion metrics stems from the requirement of carrying out measurements at full spatial scale. Table 9.1 highlights that the average SAM of the images obtained from misregistered data is simply achieved by adding the same constant to offset the corresponding distortions in the case of null shift. Conversely, ATW yields D_S 12% lower than GSA without misregistration; but when the data are shifted GSA attains a distortion that is almost 30% lower than ATW. Such distortion values match the higher degree of shift-tolerance of GSA, which has been visually remarked in Figures 9.5 and 9.6.

Eventually, fusion simulations at the degraded scale have been carried out on IKONOS data, only for GSA and ATW methods. Also in this case, the quality assessments have been performed by considering SAM, ERGAS, and Q4 indices. Figure 9.7 reports the values of the indices, measured between fused products and reference originals, spatially aligned to the degraded Pan used for fusion. Indices are plotted versus displacements between MS and Pan, which are measured in meters. Therefore, a shift equal to 32 m corresponds to two pixels at the scale of degraded MS (16 m). In this way, subpixel displacements, which occur in most practical cases, are easily assessed. Figure 9.7 provides clear evidence that misregistration always degrades the fusion performances of all methods, even though the sensitivity of GSA is appreciably lower. It is noteworthy that the loss of quality originated by misregistration is fully captured by Q4, which also exploits correlation measurements, and is particularly capable of detecting shifts. However, all indices are in trend with one another and with the visual appearance of the fused images.

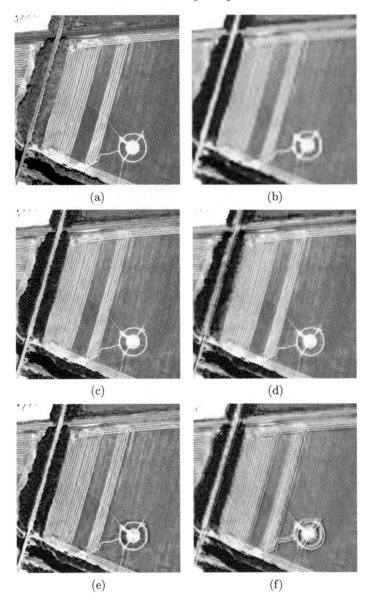

(a) (b)

(c) (d)

(e) (f)

FIGURE 9.5: Details of original and fused IKONOS data (256 × 256 at 1 m scale) showing a rural landscape. (a): Pan; (b): true-color display of 4 m interpolated MS; (c): enhanced Gram-Schmidt (GSA); (d): GSA with data misaligned by 4 pels at the Pan scale (1 pel at MS scale); (e): ATW with an MTF-matched filter and global injection model; (f): ATW with an MTF filter and global injection model with data misaligned by 4 pels at the Pan scale.

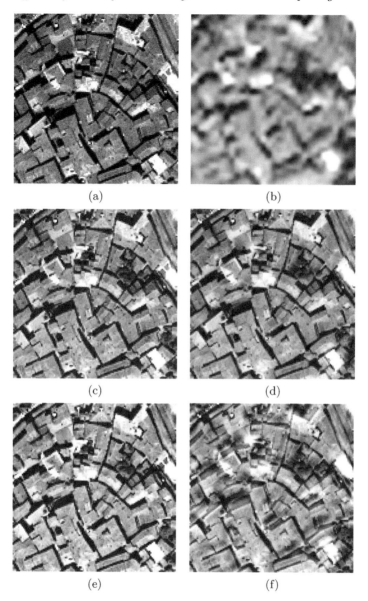

(a)   (b)

(c)   (d)

(e)   (f)

FIGURE 9.6: Details of original and fused IKONOS data (256×256 at 1 m scale) showing a little town with roofs. (a): Pan; (b): true-color display of 4 m interpolated MS; (c): enhanced Gram-Schmidt (GSA); (d): GSA with data misaligned by 4 pels at the Pan scale (1 pel at MS scale); (e): ATW with an MTF-matched filter and global injection model; (f): ATW with an MTF filter and global injection model with data misaligned by 4 pels at the Pan scale.

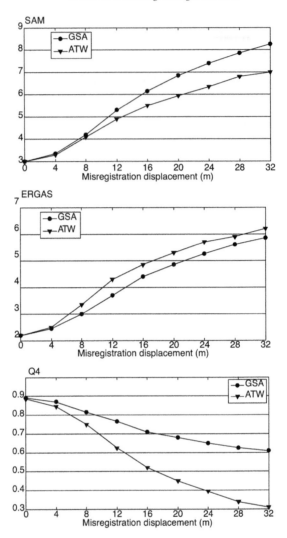

FIGURE 9.7: Quality/distortion indices for increasing amounts of misregistration between MS and Pan images. Misregistration is measured in meters (32 m = 2 pels for IKONOS MS data at a degraded spatial scale).

9.4.3.2 Interpolation shifts

In this section, the behavior of CS-based methods, for example, GSA [26], and of MRA-based methods, for example, featuring ATW with an MTF-like filter for each MS channel [14], is investigated in the presence of misalignments between MS and Pan introduced by incorrect interpolation of the former. As shown in Chapter 4, interpolation shifts can arise in the first step of most fusion processes involving data acquired by modern spaceborne sensors (IKONOS, QuickBird, GeoEye, WorldView, etc.) because of the intrinsic shift generated in the collection of MS and Pan data. Consequently, if the interpolation of the MS bands to the Pan scale is performed by using an odd filter, which intrinsically does not generate shifts in its outputs, the original shift survives and affects the fusion products. Conversely, if an even filter is adopted, the shifts generated by this type of filter can compensate the original acquisition shifts. Alternatively, if the original MS and Pan data do not have any shift, as the simulated Pleiadés data, even filters generate shifts which can affect the fusion procedure, differently from odd filters.

However, in both cases in which an interpolation shift is present, the CS-based methods are capable of mitigating the effects of the misalignments to a higher extent than MRA-based methods do, according to the behavior shown in Section 9.4.3.1 for misregistration. To demonstrate this, the simulated Pleiadés data have been considered for fusion, in two distinct tests. In the former, the original version of the Pleiadés data has been adopted, where the 3.2 m MS is odd-aligned to the 80 cm Pan, so that the performances of the two methods can be evaluated when an even interpolation filter generates a shift. This is, however, an unrealistic case in real fusion frameworks. In the latter, a new version of the Pleiadés data has been generated, where the MS bands are even-aligned to the Pan, as the alignment featured by modern instruments. In this case, the comparison is devoted to show the differences when an odd filter is used for the interpolation of the MS data.

The goal of simulations is twofold. On one side, to highlight the different degree by which the performances of the two methods are affected by an interpolation that introduces misalignments between expanded MS and Pan data. On the other side, to evidence the trend in performance index values (Q4, ERGAS, and SAM) with the order of interpolation, or equivalently with the degree of the local interpolating polynomial. Odd implementation of degrees 1, 3, 7, and 11 have been considered. Even implementations are limited to degrees 0 (NN interpolator), 1, 2, and 3, mainly because algorithms of higher degrees are presently unavailable.

Figures 9.8 and 9.9 report SAM, ERGAS, and Q4 for the two fusion methods (ATW and GSA) and for the different alignments of the input dataset, that is, odd and even alignments, respectively, between MS and Pan data, varying with the degree of interpolating polynomials and for odd and even implementation of the filters. As previously noted, only odd filters with odd-aligned data and even filters with even-aligned data produce perfectly overlapped Pan and MS interpolated by four. Instead, the other two combinations (even filters, odd data; odd filters, even data) introduce a diagonal shift by 1.5 pixels, which affects the performances of one class of filters, odd or even, with respect to the other, in the fusion procedures.

Concerning the degree of the polynomials, for odd filters, whenever suitable, fusion performances steadily increase with the order of the interpolating polynomial, in particular about two-thirds of the performance increment between linear and 11th order interpolation is achieved by the cubic kernel, which deservedly represents the best trade-off. For even filters, the trend is not monotonous, at least for SAM and Q4: NN interpolation is preferable to the even linear interpolation, which yields the worst results. Moreover, the quadratic interpolator outperforms NN and linear interpolators, but the cubic kernel is always superior.

Both in a case of an odd MS-Pan alignment, which is displayed in Figure 9.8, and in a case of an even one, which is shown in Figure 9.9, differences in ERGAS and Q4 values appear strongly mitigated in GSA fusion tests with respect to ATW, if the type of the interpolation filter is decoupled with the MS-Pan alignment itself, whereas differences in SAM are slightly emphasized. The natural explanation is that CS methods, as GSA is, are rarely sensitive to small misalignments [45], which cause progressive fading of geometry in the scene, and hence loss of performances, in MRA methods, as ATW is. Both ATW and GSA, however, cannot avoid the spatial shifts of colors caused by MS to Pan misalignment. SAM is adequate for capturing changes in color hues, but is less powerful for detecting geometric changes. This effect is visible in Figures 9.10 and 9.11, where images before and after fusion, together with a high-resolution reference original are displayed. The scale has been deliberately set in order to appreciate small shifts (1.5 pixels). That is the reason high-resolution MS and Pan originals and all fusion products are moderately sharp. The correct interpolation is with a filter of even length. Bicubic interpolation, either odd or even, has been used throughout. ATW and GSA fusion products are comparable when interpolation is even, but the former is somewhat poorer than the latter when interpolation is odd, that is, not correct.

9.5   Sensitivity to Temporal Misalignments

Assume two observations of the same scene are available at two different times; the first constituted of an MS and a Pan image; the second, more recent, made of only an MS or a Pan image. The objective is to study pansharpening for the second date with the available data. In this case, temporal misalignments can arise in the fusion processes, because the data to be merged have not been acquired in the same moment and changes in the scene have usually occurred.

In this analysis, as usual, fusion will be performed by (9.2) (for CS-based methods) or (9.4) (for MRA-based algorithms). When both MS images are available, the MS term will be up to date, while the detail term will be obtained with the Pan of the first observation. For CS-based methods, (9.2) shows that the details will be obtained as the differences of terms that are temporally misaligned because of changes occurring between the two obser-

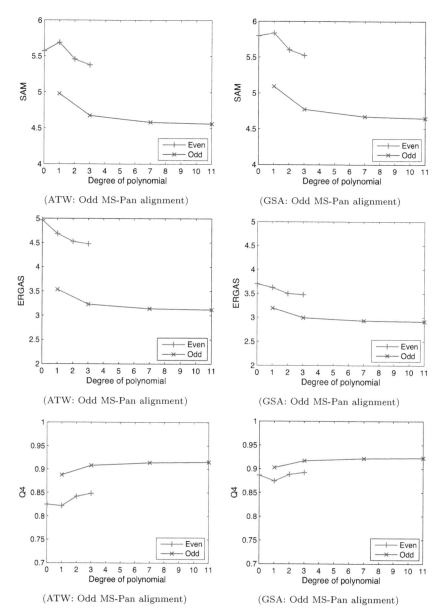

FIGURE 9.8: Quality (Q4) and distortion (SAM, ERGAS) indices of an MRA-based method (ATW) and of a CS-based method (GSA) for an odd alignment of MS to Pan and for polynomial kernels of odd and even lengths, as a function of the degree of the interpolating polynomial.

vations. Conversely, for MRA-based algorithms, (9.4) is characterized by no

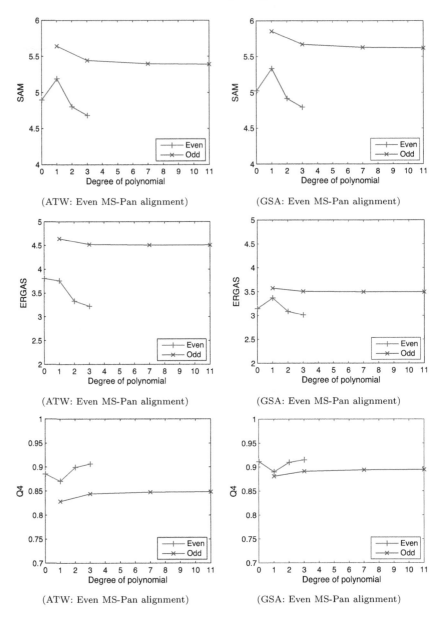

FIGURE 9.9: Quality (Q4) and distortion (SAM, ERGAS) indices of an MRA-based method (ATW) and of a CS-based method (GSA) for an even alignment of MS to Pan and for polynomial kernels of odd and even lengths, as a function of the degree of the interpolating polynomial.

temporal misalignment for details, apart they refer to the first observation

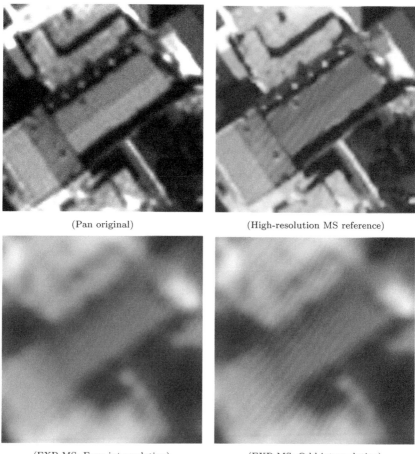

(Pan original)	(High-resolution MS reference)
(EXP MS; Even interpolation)	(EXP MS; Odd interpolation)

FIGURE 9.10: Pan, MS high-resolution reference and true-color composition of MS details interpolated by means of even- and odd-length cubic filters. The MS reference is exactly superimposed to Pan. Original MS data before interpolation are even-aligned to Pan. Interpolation with an even filter of original MS data yields no shifts between MS and Pan. Interpolation with an odd filter introduces an 1.5 pixel diagonal shift between MS and Pan. The size of all icons is 64 × 64.

and some spatial impairment could occur on edges and textures but not on homogeneous areas. As a consequence, CS techniques will exhibit stronger spectral distortion than MRA.

When only the first MS and both the Pan observations are available, the temporal misalignment of detail terms in (9.2) will be still present. In addition, the MS will forcedly be relative to the first observation and this will be the cause of severe spectral distortions for both CS and MRA schemes.

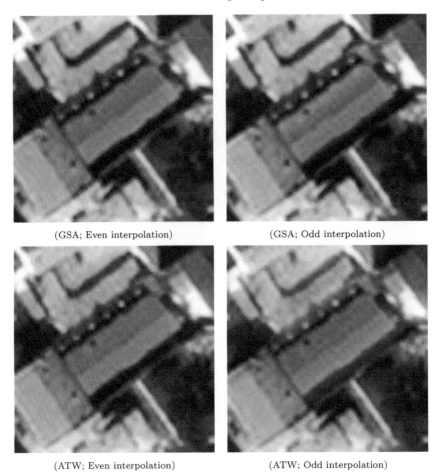

FIGURE 9.11: Fusion products of ATW and GSA starting from odd and even bicubically interpolated images displayed in Figure 9.10. By comparing with Figure 9.10, it is apparent that the incorrect odd interpolator causes color shifts, mainly visible on the roof, which appear mitigated by GSA fusion. The size of all icons is 64 × 64.

9.5.1   Results and Discussion

In this test, two GeoEye observations of the same scene have been acquired in 2010 at different times, that is, on May 27th and July 13th, respectively, with different incidence angles on the area of Collazzone, in Central Italy. The nominal spatial resolution is 0.5 m for the Pan and 2 m for the MS images, respectively. The radiometric resolution is 11 bits. All the images have been othonormalized by using a digital elevation model (DEM) available at 10 m resolution for all x-y-z spatial coordinates. Some residual spatial misalignment

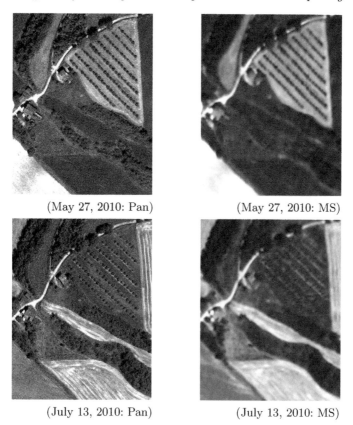

(May 27, 2010: Pan)   (May 27, 2010: MS)

(July 13, 2010: Pan)   (July 13, 2010: MS)

FIGURE 9.12: GeoEye-1 MS+Pan data (512 × 512 at 0.5 m scale). Pan and MS images acquired both on May 27, 2010; Pan and MS images acquired both on July 13, 2010.

is present after the othonormalization procedure because of inadequate resolution of the DEM. However, the presence of misalignments does not influence this analysis.

The objective is to study pansharpening for the second date of July, by comparing the case in which all the data (MS and Pan) are available and the case in which only the MS or only the Pan are available, and the other image has to be taken by the first date of May.

Figure 9.12 shows a 512 × 512 detail of the whole scene that has been processed. True-color display of MS and Pan is reported for the two dates in May and July. Many changes occurred between the two acquisitions. Such changes are of interest for the present analysis, whose main objective is investigating the behaviors of MRA- and CS-based fusion methods when temporal variations of the scene occur between the MS and Pan acquisitions. To this purpose, quantitative indices and visual evaluations have been considered. In order to

TABLE 9.2: Fusion quality/distortion scores for the MRA-based method obtained on degraded data according to Wald's protocol.

Reference = July	EXP	MRA: MS/Pan		
Score	July	July/July	July/May	May/July
ERGAS	1.91	2.16	1.90	11.68
SAM	1.61	2.19	1.67	11.86
Q4	0.898	0.911	0.896	0.323
Reference = May				
Score	July	July/July	July/May	May/July
ERGAS	10.31	10.51	10.19	1.85
SAM	11.98	12.13	12.00	2.31
Q4	0.420	0.406	0.456	0.876

TABLE 9.3: Fusion quality/distortion scores for the CS-based method obtained on degraded data according to Wald's protocol.

Reference = July	EXP	CS: MS/Pan		
Score	July	July/July	July/May	May/July
ERGAS	1.91	2.18	3.71	10.49
SAM	1.61	2.18	2.15	11.57
Q4	0.898	0.908	0.731	0.453
Reference = May				
Score	July	July/July	July/May	May/July
ERGAS	10.31	10.52	9.70	6.15
SAM	11.98	12.13	12.11	3.33
Q4	0.420	0.403	0.565	0.586

produce performance values, all the data have been degraded according to Wald's protocol by a scale factor equal to four. Fusion has been accomplished according to three different modalities: (i) with both MS and Pan images of July; (ii) with MS of July and Pan of May; (iii) with MS of May and Pan of July. The two simple yet optimized methods developed by the authors and denoted as GLP and GSA in [25] have been selected as representative of MRA- and CS-based fusion.

Tables 9.2 and 9.3 report values of ERGAS, SAM, and Q4 indices in the three cases: (i), (ii), and (iii), for MRA-based and CS-based methods, respectively. Also the scores of void fusion without injection of details, have been reported under the entry EXP. Performances obviously depend on the MS image chosen as a reference, which represents the target of fusion. In the upper

(MRA fusion: MS=July+Pan=July)

(MRA fusion: MS=July+Pan=May) (MRA fusion: MS=May+Pan=July)

FIGURE 9.13: Details of MRA-fused GeoEye-1 data (512×512 at a 0.5 m scale), by considering both MS and Pan data at the same time, or by fusing MS and Pan images taken at different times, as reported in Figure 9.12.

tabular, the indices are calculated with the original MS image of July as reference. In the lower tabular, the reference is the original May MS image. When the reference is the July MS image, in case (i) MRA and CS exhibit almost identical scores. In case (ii) MRA obtains a surprising result since ERGAS and SAM are somewhat lower than the (i) case; conversely, Q4 is lower than before, thus revealing that the quality of the injected details is good enough but not as good as in case (i); results of CS reveal a steady decrement in performances. The scores of case (iii) are undoubtedly poor; it is noteworthy that CS is better than MRA, but this result was expected. In fact, as a general consideration, MRA produces images that are spectrally more similar to the MS images than CS does; conversely, CS yields fusion products that are spatially more similar to the Pan image than MRA does. The lower tabular substantially shows that, when the May image serves as reference, only MRA fusion with May MS and July Pan is acceptable. The second best result is CS

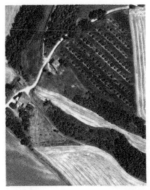

(CS fusion: MS = July+Pan = July)

(CS fusion: MS = July+Pan = May) (CS fusion: MS = May+Pan = July)

FIGURE 9.14: Details of CS-fused GeoEye-1 data (512 × 512 at a 0.5 m scale), by considering both MS and Pan data at the same time, or by fusing MS and Pan images taken at different times, as reported in Figure 9.12.

with May MS and July Pan. Interestingly, the third best result is when the high-resolution May MS image is synthesized by CS through the July MS and the May Pan images.

Figures 9.13 and 9.14 show the details of the images whose scores are reported in Tables 9.2 and 9.3 for the MRA- and CS-based methods, respectively. The upper parts of Figures 9.13 and 9.14 refer to case (i). Visual analysis reveals that the quality of the fused products is very good and quite comparable. The middle parts of Figures 9.13 and 9.14 refer to case (ii). MRA maintains most of the color feature of the original MS image and only some blur due to temporal changes and to misregistration appears. Conversely, CS exhibits a heavy color distortion even if spatial details are sharper than for MRA. Eventually, the images fused according to case (iii) are shown in the lower parts of Figures 9.13 and 9.14. Photo analysis reveals that spectral quality is poor. However, spatial details of CS are sharper than those of MRA.

9.6 Concluding Remarks

In this chapter, the sensitivities of the two main families of pansharpening approaches to different types of distortions, namely aliasing and both spatial and temporal misalignments have been investigated. Table 9.4 recalls the conclusions given by the tests that have been discussed in the previous sections. In particular, CS-based methods are optimal to cope with aliasing distortion and spatial misalignments between original MS and Pan data, both caused by misregistrations in the data acquisition, or unsuitable interpolation in the first step of the fusion procedure. Conversely, MRA-based methods are preferable in the case of temporal misalignments that can arise when the processed images have been acquired on different times. Among MRA-based methods, GLP is capable of partially recovering aliasing distortion, even if to a lesser extent than CS-based algorithms.

TABLE 9.4: Sensitivities of the main families of pansharpening methods to various types of distortions.

	Aliasing	Space Shift		Time Shift
		Misreg.	Interp.	
CS	Low	Low/Med.	Low/Med.	High
MRA-GLP	Low/Med.	High	High	Low/Med.
MRA-ATW	High	High	High	Low/Med.

Chapter 10

Fusion of Images from Heterogeneous Sensors

10.1 Introduction

The term *heterogeneous* sensors indicates instruments that do not exploit overlapped wavelength intervals and/or same imaging mechanisms to produce observations. Fusion of images acquired by *heterogeneous* sensors is gaining an ever increasing relevance in application fields where, notwithstanding technological progresses and developments, intrinsic limitations of imaging sensors cannot offer substantial benefits in a unique dataset.

This scenario also occurs in the remote sensing field because of the heterogeneous nature of sensors that are currently employed for a variety of applications, as discussed in Chapter 2, where remote sensing principles were introduced and basic concepts on sensors were presented. In particular, passive and active sensors exist that operate in a wide range of spectral wavelengths.

While in pansharpening, either MS or HS, the fusion process is limited to the wavelengths of the solar radiation reflected by the Earth's surface, the possibility of sharpening a thermal image, whose pixel values are proportional to the radiation emitted by the Earth's surface because of its absolute temperature, requires the availability of a more spatially resolved image. An image with such characteristics is not attainable by any thermal imaging instrument, but is easily obtained from a sensor imaging the Earth's surface in the visible and/or NIR portions of the spectrum (V-NIR).

In the following sections, we present two cases involving fusion of images acquired from heterogeneous sensors:

- spatial enhancement of thermal image(s) with higher-resolution V-NIR observations;

- enhancement of optical images by means of SAR observations of the same scene;

and present as many examples of specific applications of fusion. In the former case, thermal image bands acquired by ASTER are sharpened by means of a visible band acquired from the same platform. In the latter case, ERS SAR images of an urban area are used to extract and inject significant spatial features into an MS Landsat 7 ETM+ image of lower-spatial resolution.

As it will be discussed in the next sections, methods like those described in Chapter 7 may not be applied straightforwardly because of the different nature of data to be fused. A proper choice or adaptation of the most suitable method among those already existing should be preliminary envisaged, in order to get fused products that retain the significance of the component datasets, from either a physical or an informative viewpoint. Also issues concerning quality evaluation of fusion methods shall be addressed, as discussed in Chapter 3, and adapted to the specific cases.

10.2 Fusion of Thermal and Optical Data

As explained in Section 2.4.3, thermal sensors acquire the radiation emitted by the Earth surface because of its temperature according to Planck's law (2.2). Since the radiant emittance of the Earth at standard environmental temperatures is low and thermal sensors are somewhat noisy, because of large dark signal fluctuations, the spatial resolution of TIR images is necessarily scarce, in order to capture a sufficiently high number of photons of the weak at-sensor radiance. In fact, the most feasible way to improve SNR is increasing the integration area, thereby reducing the spatial resolution. Hence, the potential utility of enhancing the spatial resolution of TIR images stands out.

Unfortunately, spatial sharpening of TIR images is not immediate since the spectral channels of TIR do not overlap with the spectral channels of higher-resolution spectral bands that might be exploited for sharpening [24]. In fact, spectral overlapping would guarantee some correlation between the sharpened and the sharpening images, like in MS pansharpening, and correlation is the prerequisite for good fusion. In principle, there is no evidence that TIR and V-NIR radiances are correlated, being the former emitted and the latter reflected by the Earth's surface and moreover in totally disjoint spectral intervals (wavelengths of TIR are approximately 10 times greater than those of V-NIR).

10.2.1 Background and Literature Review

To overcome the lack of spectral overlap, many authors have addressed the problem of spatial enhancement of TIR data by exploiting some physical relationships between the TIR band(s) and other bands having higher-spatial information. A notable example is constituted by the relationship observed between vegetation and temperature [192]. The idea is to calculate the *vegetation index* (VI) or another feature derived from it, like the *normalized difference vegetation index* (NDVI), as a scaling factor to predict thermal images at higher-spatial resolutions. In fact, NDVI is computed by utilizing the Red and V-NIR bands that are usually available at a better spatial resolution than TIR. The idea, originally reported by Moran [192], was later developed in Kustas *et al.* [156], further improved in Agam *et al.* [1], and revisited in Jeganathan [142] and Chen *et al.* [69].

In a very recent paper [274], the problem of TIR sharpening is formulated in terms of *disaggregation* of remotely sensed *land surface temperature* (DLST) and two subtopics, *thermal sharpening* (TSP) and *temperature unmixing* (TUM), are identified as a dual pair of DLST. TSP refers to any procedure, through which thermal images are enhanced for the purpose of interpretation by using spatially distributed auxiliary data that are statistically correlated to LSTs, pixel by pixel, block by block, or region by region. Instead, TUM denotes the procedures, by which component temperatures within a pixel are decomposed on the basis of multitemporal, spatial, spectral, or angular observations [274].

Several fundamental assumptions are usually adopted in DLST:

1. resolution cells (pixels) are larger than image objects and thus the surface should be thermally heterogeneous at the pixel scale;

2. at the object scale, the surface is assumed isothermal (separability) [121];

3. auxiliary data relating to the physical properties of subpixel elements are available (connectivity);

4. aggregation occurring among objects is additive (additivity) [121].

The first assumption implies that a pixel consists of an array of components that characterize the mixture among multiple elements. The second is unrealistic but is assumed as true in most of previous studies. The third assumption is necessary to keep the number of unknowns as small as possible because such auxiliary data can be used to determine the fractions of elements or to estimate the physical properties of elements. The last assumption is mostly reasonable when there is no energy interaction among components, but the aggregation is generally nonlinear when heterogeneous surfaces have a significant three-dimensional structure, much larger than the measuring wavelength, which may cause inter-shadows and thus create nonlinear interactions between sub-facets [183]. An extensive treatment of DLST, however, is beyond the scope of this book. For an in-depth insight on DLST methods and for a thorough collection of references, the reader can refer to the review paper by Zhan *et al.* [274].

Upon these premises, the following section introduces a feasible approach that can be adopted for sharpening TIR data. Such a method belongs to the TSP family and exploits the presence of a certain correlation existing between TIR and some of the other higher-resolution bands. Originally proposed in Aiazzi *et al.* [24], the method is substantially a classic MRA pansharpening algorithm. Due to its interesting spectral characteristics, the presence of multiple TIR bands, a specific case study concerning the ASTER sensor will be presented and discussed. Extensions to different instruments would be quite immediate.

10.2.2   Fusion of ASTER Data

ASTER, an acronym for *Advanced Spaceborne Thermal Emission and Reflection Radiometer*, is a Japanese sensor, which is one of five RS instruments onboard the Terra satellite launched into Earth's orbit by NASA in 1999. The instrument has been collecting superficial data since February 2000.

ASTER provides high-resolution images of our planet in 14 different bands of the electromagnetic spectrum, ranging from visible to thermal infrared light. The resolution of images ranges between 15 m and 90 m. ASTER data are used to create detailed maps of surface temperature of land, emissivity, reflectance, and elevation. Unfortunately, notwithstanding the global Earth acquisition capability of the instrument, analogous to Landsat (E)TM(+), the lack of a band in the blue wavelengths prevents true-color pansharpened products (Google Earth or similar) from being achieved.

Before introducing the problem of fusing ASTER data, a brief description of the ASTER imaging instrument is reported. Then, the specific adaptation of the GLP-HPM algorithm to ASTER data will be presented. The problem of correlation between the TIR bands and the V-NIR bands, from which the HR spatial details will be extracted will be addressed. Ultimately, some examples of fusion will be presented, to give an idea of the results that can be achieved and to highlight some key points that will be the object of a brief discussion.

TABLE 10.1: Characteristics of ASTER imaging sensor. Spatial resolution decreases as wavelength increases. Radiometric resolution is 8 bit/pel for V-NIR and SWIR bands and increases to 12 bit/pel for TIR bands.

V-NIR: 15 m - 8 bit l	SWIR: 30 m - 8 bit	TIR: 90 m - 12 bit
B1: 0.52 - 0.60 μm	B4: 1.600 - 1.700 μm	B10: 8.125 - 8.475 μm
B2: 0.63 - 0.69 μm	B5: 2.145 - 2.185 μm	B11: 8.475 - 8.825 μm
B3: 0.76 - 0.86 μm	B6: 2.185 - 2.225 μm	B12: 8.925 - 9.275 μm
	B7: 2.235 - 2.285 μm	B13: 10.25 - 10.95 μm
	B8: 2.295 - 2.365 μm	B14: 10.95 - 11.65 μm
	B9: 2.360 - 2.430 μm	

The images that were processed have been extracted from the sample L1B data of ASTER imagery of Mt. Fuji made available by the *Earth Remote Sensing Data Analysis Center* (ERSDAC) of the Ministry of Economy, Trade and Industry (METI) of Japan.

10.2.2.1 ASTER imaging sensor

In order to devise a viable strategy for enhancing the spatial resolution of the ASTER TIR data let us first examine the characteristics of the ASTER sensor reported in Table 10.1.

There are three groups of bands in three different spectral ranges, V-NIR, SWIR, and TIR, with a different spatial resolution. The strategy that appears as the best would be utilizing some of the V-NIR bands with a resolution of 15 m for sharpening the TIR bands, which have a resolution of 90 m. The 1:6 scale ratio between V-NIR and TIR bands is perhaps a limit case for the feasibility of pansharpening. A further consideration concerns TIR bands. Such bands usually exhibit an extremely high degree of correlation between one another. Presumably, there will be a unique V-NIR band featuring the highest degrees of correlation with all the TIR bands. Once this band has been detected, V-NIR-sharpening of TIR bands can be accomplished. If such a correlation is sufficiently high, the results of fusion will most likely be acceptable.

10.2.2.2 A Fusion scheme for TIR ASTER data

Let us now present a simple fusion scheme that can be applied to ASTER data. Because of its characteristics, an MRA algorithm with the HPM model (7.10) appears to be the best solution. In fact, if the multiband image pixel is represented as a vector, the detail by which it will be enhanced is parallel to it. This is the most immediate way to preserve the spectral characteristics of the multiband image. In the case of ASTER TIR fusion, the role of the MS image is played by the TIR image, whose pixels may be regarded as vectors with five components (bands 10, 11, 12, 13, and 14). The spectral preserving nature of the HPM model is fundamental for TIR fusion: because of the different nature of TIR and V-NIR radiances, HPM can guarantee that, apart from the injection of highpass spatial detail, no further spectral distortion is introduced in the TIR because of the different nature of V-NIR radiance. HPM is also capable of taking into account the different radiometric resolution of the V-NIR and TIR bands. Eq. (7.10), rewritten for the present case, becomes

$$\widehat{T}_k = \widetilde{T}_k \left(1 + \frac{V - V_L}{V_L} \right) = \widetilde{T}_k \cdot \left(\frac{V}{V_L} \right) \tag{10.1}$$

in which T_k denotes the kth TIR band, $\sim$ and $\wedge$ indicate expansion, that is, plain interpolation, and spatial enhancement, that is, interpolation and addition of details, respectively. V is the enhancing V-NIR band from which the detail is extracted and V_L its lowpass-filtered version, if the type of MRA used in this context is ATW. One can easily note that the spatially modulating factor on the right side of (10.1) is independent of the different radiometric resolutions of V-NIR and TIR.

Figure 10.1 shows the MRA fusion scheme with the HPM injection model, denoted in the following as GLP-HPM, which has been adjusted for V-NIR and TIR ASTER data. The 6:1-reduced V-NIR image is obtained by means of the cascade of two reduction steps ($p_1 = 3$, $p_2 = 2$, respectively). The 1:6 expansion of the reduced V-NIR and TIR images are achieved through two cascaded expansion steps ($p_1 = 2$ and $p_2 = 3$). The injection model is computed at the scale of the fusion product between the lowpass approximation of V-NIR and the expanded TIR bands. $\downarrow (\cdot)$ and $\uparrow (\cdot)$ denote downsampling and upsampling. $r_{(\cdot)}$ and $e_{(\cdot)}$ denote reduction and expansion. Reduction filters are roughly Gaussian with amplitude at Nyquist equal to 0.25. Interpolation kernels by two and three are those described in Tables 4.3 and 4.4. Note that if the underlying MRA is (partially) decimated, like GLP, and not undecimated, like ATW, V_L should be replaced by $V_L^* = ((V_L \downarrow r) \uparrow r)_L$, that is, V_L^* is the lowpass-filtered version of $(V_L \downarrow r) \uparrow r$ (see Section 9.3.2.2).

10.2.2.3 Interchannel correlation

As previously pointed out, the statistical similarity of the images to be merged is crucial for the quality of fusion. Table 10.2 details the correlation matrix, calculated between the original TIR bands (10 through 14) and V-NIR bands (1, 2, and 3), spatially degraded by a factor six.

An analysis of Table 10.2 reveals that V-NIR band 2, corresponding to the wavelengths of red color, strongly correlate with all the thermal bands. Thus, it is the most suitable image among the three V-NIR bands for acting as an enhancing Pan, that is, for originating the spatial details to be injected in the expanded TIR bands. In addition, as expected, all the TIR bands are extremely correlated to each other. Visual analysis of Figure 10.2 confirms that, apart from some difference in contrast, the thermal band 10 is very similar to band 2 (red) while strong differences exist with respect to band 3 (NIR). The same conclusion is obtained when a visual analysis is performed among bands 2 and 3 and bands 11, 12, 13, 14 (TIR).

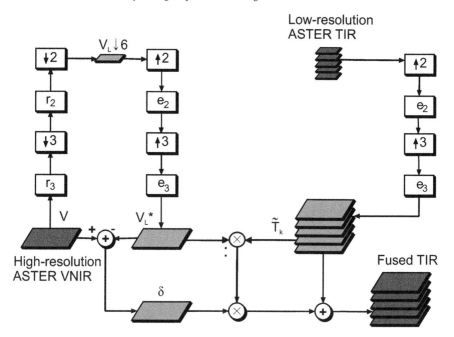

FIGURE 10.1: GLP-based pansharpening for the fusion of ASTER TIR bands with one V-NIR band. The scheme is a GLP (Section 5.5.1), in which the reduction by $p = 6$ is achieved as the cascade of two reductions by $p_1 = 3$ and $p_2 = 2$. Analogously, the expansion by $p = 6$ is obtained as the cascade of two expansions by $p_1 = 2$ and by $p_2 = 3$, respectively. The injection model (10.1) is calculated at the scale of the fused product between the lowpass approximation of V-NIR and the expanded TIR bands.

TABLE 10.2: Correlation between V-NIR and TIR ASTER bands for the test image. Values above the main diagonal are calculated on the whole images of Figure 10.2; values below the diagonal refer to the upper left quarter of Figure 10.2(a) to avoid the influence of the sea, whose effect is to increase correlation.

Band	1	2	3	10	11	12	13	14
1	1	0.883	0.064	0.663	0.662	0.658	0.673	0.676
2	0.935	1	0.442	0.891	0.891	0.887	0.891	0.892
3	-0.518	-0.263	1	0.550	0.556	0.557	0.531	0.523
10	0.711	0.858	0.027	1	0.995	0.993	0.993	0.992
11	0.709	0.856	0.035	0.994	1	0.995	0.994	0.991
12	0.704	0.851	0.044	0.993	0.995	1	0.993	0.991
13	0.725	0.864	0.011	0.993	0.994	0.994	1	0.997
14	0.732	0.870	-0.002	0.991	0.991	0.991	0.997	1

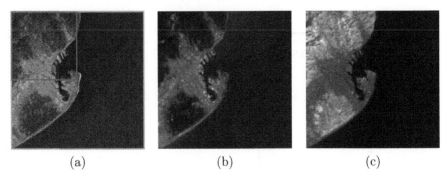

FIGURE 10.2: (a): TIR B10; (b): V-NIR B2; (c): V-NIR B3. V-NIR bands have been reduced to the same scale of TIR by proper lowpass filtering and decimation by six.

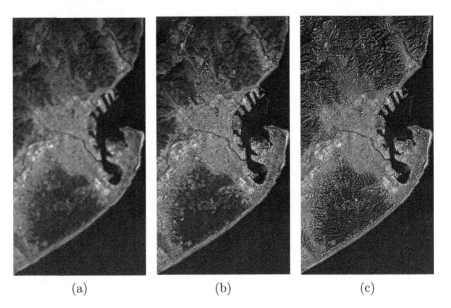

FIGURE 10.3: False color-composite pictures of TIR bands 14, 12, and 10 displayed as R, G, and B channels: (a): expanded TIR, TIR fused with V-NIR; (b): band 2; (c): band 3.

10.2.2.4   Results and discussion

To quantify the concept that similarity between images to be fused is crucial and that HPM is suitable for avoiding spectral distortions, two simulations are presented and discussed in parallel. In the former, spatial details are extracted from the highly correlated band 2 and merged according to the scheme outlined in Figure 10.1. In the latter, the sole difference is that details are extracted from the weakly correlated band 3.

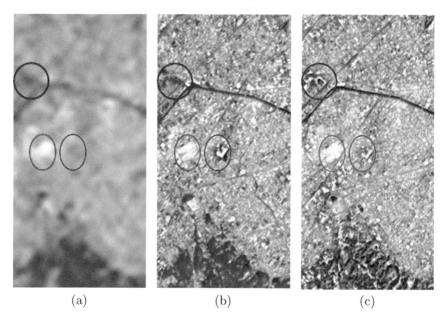

(a)   (b)   (c)

FIGURE 10.4: Details of the image in Figure 10.3. Objects in the TIR image (a) appear sharpened in the images fused with (b) band 2 and (c) band 3.

The fusion results are reported in Figure 10.3 on a subset of the TIR bands. Bands 14, 12, and 10 are mapped in a false-color composition onto the R, G, and B channels, respectively. The original 90 m TIR image (Figure 10.3(a)) has been expanded to the same scale of V-NIR bands and can be compared with the TIR images fused by means of V-NIR band 2 (Figure 10.3(b)) and 3 (Figure 10.3(c)). As expected from the correlation analysis, the original TIR image is more similar to the image enhanced by band 2. In particular, fusion by band 3 suffers from an overenhancement of vegetated areas, which appear extremely textured, because of the strong response of vegetation in the NIR channel. Although different in contrast, both the fused images appear spectrally similar, that is, they exhibit no color inaccuracies, as otherwise expected from the angle-preserving property of the HPM model.

Details of the image in Figure 10.3 are shown in Figure 10.4. Visual analysis demonstrates that objects in the TIR image (Figure 10.4(a)) appear sharpened in the fused images (Figures 10.4(b) and (c)). The scarce correlation existing between the TIR bands and band 3, makes the details introduced by fusion with band 2 to be more consistent than those of the fusion with band 3. The overall effect of TIR+red fusion is impressive and the sharpened images seem to be a valid tool for photo analysis.

A second detail extracted from the image in Figure 10.3 portrays a vegetated area shown in Figure 10.5. The sharpened image produced by extracting the injected details from band 3 (Figure 10.5(c)) appears overenhanced and

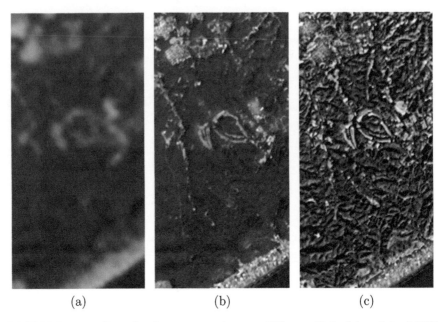

(a)   (b)   (c)

FIGURE 10.5: Details of a vegetated area of Figure 10.3: (a): original TIR bands; (b): TIR bands merged with band 2; (c): TIR bands merged with band 3.

poorly correlated with the original TIR image (Figure 10.5(a)). Conversely, the image fused by means of band 2 (Figure 10.5(b)) is spectrally similar to the (expanded) original and the spatial details that have been injected appear most likely.

Concerning quantitative results, the consistency property of Wald's protocol has been checked: for each band, the fused image is reduced to its native resolution and compared with the original image. Consistency requires that the two images are ideally equal. Differences are assessed by such distortion measures as mean bias, standard deviation bias, and mean square error (MSE). The value of these distortions is required to be as low as possible, ideally zero. Similarily, the measured correlation coefficients (CC) should ideally be equal to one.

The results are reported in Tables 10.3 and 10.4 for the fusion with band 2 and band 3, respectively. The bias in the mean is practically zero for both fused images, since mean conservation is an intrinsic property of MRA fusion algorithms. Concerning the other indices, the results reported in Table 10.3 are better than those reported in Table 10.4. For each band, the difference in standard deviation and the MSE are lower, while the CC is higher, in Table 10.3 than in Table 10.4. This happens without exceptions for all the TIR bands.

Also the spectral angle mapper (SAM) has been computed to assess the

TABLE 10.3: Score parameters of TIR bands fused by means of V-NIR band 2.

TIR band	Mean Bias	St. Dev. Bias	MSE	CC
10	0.0367	1.9652	135.4188	0.9789
11	0.0364	2.0378	154.9750	0.9800
12	0.0394	2.2453	172.1465	0.9774
13	0.0079	2.6443	239.3021	0.9745
14	0.0530	2.7816	251.5762	0.9703

TABLE 10.4: Score parameters of TIR bands fused by means of V-NIR band 3.

TIR band	Mean Bias	St. Dev. Bias	MSE	CC
10	0.0047	2.8453	276.3941	0.9576
11	0.0062	2.9680	314.3599	0.9601
12	0.0063	3.3885	352.2002	0.9548
13	0.0086	4.3424	512.3738	0.9471
14	0.0271	4.8563	559.6409	0.9368

spectral distortion of the fusion process. As expected, it was found to be practically zero in both cases (0.0334 and 0.0333 degree, respectively) because of the intrinsic spectral properties of the HPM model. It is important to evidence that the scores reported in Tables 10.3 and 10.4 are absolute values and that they are rather low, notwithstanding that the dynamic range of the TIR images is 12 bit.

10.3   Fusion of Optical and SAR Data

The integration of optical and SAR data is a challenging research topic. The two sensors operate in totally different modalities. An optical sensor acquires the solar radiation reflected by the Earth's surface and is therefore strongly influenced by environmental conditions. A SAR system records the radiation backscattered from the surface hit by the electromagnetic signal irradiated by the system itself. Thus, a SAR system is able to operate at nighttime and in all weather conditions, because the cloud cover is transparent to microwaves.

Apart from the acquisition geometry that affects the processing, the interpretation of the data is substantially different. The information contained in optical images depends on the spectral reflectance of the target illuminated by sunlight, whereas SAR image reflectivity mainly depends on the dielectric and geometric surface characteristics of the object, such as roughness (compared to the wavelength) and moisture, that is, water content, as well as on frequency, polarization, and incidence angle of the electromagnetic radiation. As a consequence, the traditional fusion methods of optical data must be adapted to the new problem or even designed anew.

10.3.1 Problem Statement

Due to the different and complementary nature of optical and SAR data, their integration is a research and technological task of great interest, but of uneasy realization. Its fulfillment will allow a series of subtargets and of scientific and technological results to be achieved. Although the solution may not be immediate, given the physical heterogeneity of the data, fusion of optical (visible, NIR, and SWIR) and SAR image data may contribute to a better understanding of the objects observed within the imaged scene [212].

First, let us recall the types of data that are currently available or will be available shortly. In particular, new generation SAR with multilook geocoded products at 5–6 m, is essential. At the same time there are several MS and Pan optical sensors at medium (Landsat 8, ASTER) and very high (IKONOS, QuickBird, GeoEye, WorldView-2, Pléiades) spatial resolutions. Other optical sensors will be launched by ESA (Sentinel-2) and, in the context of ASI and DLR hyperspectral missions, PRISMA and EnMAP, respectively.

The main scientific goals of optical and SAR image fusion are the following:

- design area-based procedures able to produce fused images suitable for both application-oriented photo analysis and for input to fully automatic or semi-automatic application procedures;

- integrate despeckling algorithms into the fusion procedure;

- design optimal filtering procedures for SAR images taking advantage of the spatial information derived through the integrated use of optical images;

- validate optical+SAR fusion products according to statistical indices measuring the consistency properties with respect to the original (despeckled) SAR image (spatial/radiometric consistency) and to the original MS data (spectral consistency);

- assess the fusion result by comparing the thematic maps obtained from fusion products to those obtained from original data without fusion.

To achieve these goals, two different approaches can be considered, depending on the relative resolutions between optical and SAR images.

The "pansharpening-like" approach developed in Alparone *et al.* [34] consists of producing an optical image enhanced by some kind of feature extracted from SAR. Whenever SAR images have more resolution than optical images, the goal is creating an image that features the spatial detail of the SAR image and the spectral information of the (MS) optical image. This modality substantially consists of "coloring" the SAR image [55], preliminarily cleaned from speckle. This typology of procedures is suitable, for example, for merging COSMO-SkyMed or TerraSAR-X 4-look images with Landsat 8 (30 m MS, 15 m Pan), or ASTER (15 m V-NIR) images.

Conversely, when the spatial resolution of the optical image is greater than that of the SAR image, it is difficult to increase the information of the optical data by means of the SAR data. However, it is reasonable to use the spatial detail information of the VHR optical image to achieve an optimal despeckling of the SAR image. Actually, in a multiresolution representation of the SAR image, in which details are sparse while the noise is spread over all coefficients it is possible to use techniques borrowed from "compressed sensing" and "sparse coding" representations [90] to optimally reject the noise with the aid of dictionaries of high-resolution spatial patterns. Such dictionaries are generally not available, but if we assume that they can be extracted somehow from the optical data, for example, from one band or from a combination of a ratio of bands, chosen in such a way that a property of interest of the SAR image is highlighted, it would be possible to produce a SAR image cleaned from speckle that has inherited some spatial and possibly also spectral features from the optical image. This approach can also be regarded as a joint despeckling and fusion of SAR and optical images.

10.3.2 Literature Review

While there is an extensive knowledge in the fusion of homogeneous data, integration of optical and SAR data is an open problem and represents an applied research topic of great interest. After proper orthorectification of all images based on the DEM of the area, and accurate coregistration at the minimum scale for geometric congruence, fusion can be tackled at three levels:

1. "pixel" level: "area-based" methods that produce a fused image, typically with spectral diversity [34], which can be analyzed by an expert or automatically;

2. "feature" level: pointwise combination of statistical or geometrical features by exploiting the local context, according to some criterion seeking for maximizing the success of a specific application [184, 213]. These methods do not usually produce an image that can be photo-interpreted, but rather a map of the specific property under investigation;

3. "decision" level: combination of decisions obtained separately on the individual images [224]. These methods do not produce a single fused image, but only classification maps.

"Area-based" methods are seldom encountered in the literature, because they are not tied to a single well-defined problem, but are potentially able to solve wider classes of problems. In Alparone *et al.* [34], a multiplicative combination (modulation) between the generalized intensity derived from the MS data and a textural characteristic extracted from the SAR data has been proposed. The procedure is designed to combine physically heterogeneous data and to adjust automatically the amount of SAR textural information to be injected into the fused image: to a larger extent where it is relevant (for example in urban areas [108]); to a smaller extent on rural areas, where the information provided by the MS data is generally more significant. This approach extends existing solutions for pansharpening of MS data [174, 159], to the integration of SAR and MS imagery. According to this approach, SAR texture modulates the generalized intensity, which is obtained by applying an invertible linear transformation to the original MS data. Indeed, texture is one of the most important features of SAR images [253]. It provides information on the morphology of the scene on the scale at which backscatter phenomena take place, and can be characterized at pixel level by multiresolution analysis.

Different approaches aimed at multisensor image fusion are based on either decision-level fusion or multisource classifiers [225, 224]. Users often aim at enhancing application relevant features in the fused product. Hence, application specific methods more than techniques of general use, as in the case of MS+Pan fusion, have been developed [190]. The challenging task of fusion of hyperspectral and SAR imagery was recently investigated with promising result [67].

The problem with speckle, the signal-dependent coherent noise affecting SAR images [42], is also not negligible, since it should be avoided directly merging high-SNR optical data with low-SNR SAR data. Area-based techniques can *enhance* the SAR image by means of the optical data, or alternatively enhance MS data by means of SAR as in [34]. The former techniques are robust to the possible unavailability of optical images, for example, in the case of partial cloud cover.

10.3.3   Quality Issues

If quality assessment of optical image fusion, like pansharpening, is an ill-posed problem, quality assessment of optical and SAR data fusion is a doubly-ill posed problem. In fact, in the former a reference of quality exists, the MS image taken by the same instrument from a height lower as many times as the scale ratio between MS and Pan datasets, but such a reference is unavailable in practice. In the latter, even the quality reference does not exist. Hence, there is no target of fusion, at least in terms of image product generated by some instrument. Therefore, quality shall be carried out *a priori*, by verifying the consistency of the fused image with its source images, as indicated in Section 3.3, and *a posteriori* through the judgment of experts and the use of automated analysis or thematic classification, either supervised or not. It is noteworthy that *a posteriori* quality measurement may be either subjective, because of humans, or unreliable, because of inaccuracy of ground truth for classification.

10.3.4 Fusion of Landsat 7 ETM+ and ERS SAR Data

An area-based method suitable for merging an optical image (MS + Pan) and a SAR image having greater resolution is reviewed here. Since its publication in 2004 [34], the method has encountered a certain popularity, being substantially a SAR-adjusted pansharpening suitable for the most general purposes.

The spatial resolution requirements of the method impose that the optical image is a Landsat 7 ETM+ image and the SAR image is from an ERS platform. Nowadays, Landsat 7 is no longer active and Landsat 8 took its place. Analogously, the updated SAR platform of comparable resolution and features in general will be Sentinel-1 (see Table 2.3 in Section 2.5.2).

10.3.4.1 Fusion scheme

Figure 10.6 illustrates the multisensor image fusion algorithm for Landsat 7 ETM+ and ERS data proposed in Alparone *et al.* [34]. The input dataset is composed of:

- an amplitude SAR image, **SAR**, geocoded and resampled at its own resolution,

- an MS image with K bands resampled and coregistered at the scale of the SAR image and thus oversampled, namely $\tilde{\mathbf{M}}_1, \tilde{\mathbf{M}}_2, \ldots, \tilde{\mathbf{M}}_K$,

- a Pan image **P**, resampled and coregistered with SAR.

The original SAR and Pan images should have spatial resolutions greater than those of the MS image.

The SAR image is filtered to mitigate speckle, while preserving texture, thus obtaining the image $\mathbf{SAR}_{ds}$. After despeckling, the ratio **t** between $\mathbf{SAR}_{ds}$ and its lowpass version, obtained by means of the ATW algorithm as an approximation at level L, is calculated. The highpass details of the Pan image (level $l = 0$ of the ATW transform of Pan) are "injected" (that is, added) to the resampled MS bands after equalization by a spatially constant gain term, calculated by matching the histograms of original intensity and lowpass Pan, $\mathbf{P}_L$.

Despeckling is a key point of the fusion procedure. An efficient filter for speckle removal is required to reduce the effect of the multiplicative noise on homogeneous areas, while point targets and especially textures must be carefully preserved. Thus, the ratio of a despeckled SAR image to its lowpass approximation, which constitutes the modulating texture signal, is practically equal to one on areas characterized by a constant backscatter coefficient (for example, on agricultural areas), while it significantly drifts from unity on highly textured regions (urban and built-up areas) [253]. In such regions, intensity modulation is particularly effective: spatial features that are detectable in the SAR image only can be properly introduced by the fusion procedure

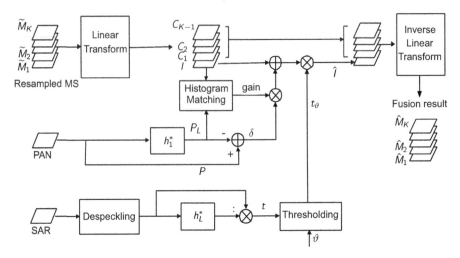

FIGURE 10.6: Flowchart of the procedure for fusion of a low-resolution K-band MS image with a high-resolution Pan image and a high-resolution SAR image.

into the VIR image, without degrading its spectral signatures and radiometric accuracy, to provide complementary information.

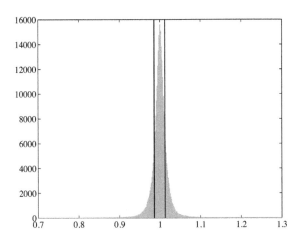

FIGURE 10.7: Histogram of modulating SAR texture $\mathbf{t}$: threshold $\theta = \sigma_t$ is highlighted.

Due to unbiasedness of despeckling performed in the undecimated wavelet domain without resorting to logarithm [105, 41, 268, 42], the ratio image $\mathbf{t} = \mathbf{SAR}_{ds}/(\mathbf{SAR}_{ds} \otimes h_L^*)$ has unity mean value, as shown in Figure 10.7. Spurious fluctuations in homogeneous areas around the mean value $\mu_t \approx 1$,

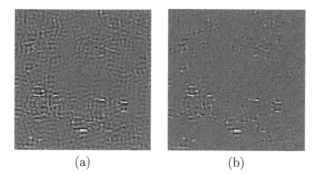

(a)   (b)

FIGURE 10.8: SAR texture (a) before and (b) after soft thresholding.

due to imperfect despeckling, intrinsic signal variability of rough surfaces, and instability of image ratioing, are clearly visible in Figure 10.8(a).

By denoting with σ_t the standard deviation of the distribution of the modulating SAR texture $\mathbf{t}$, a soft thresholding with threshold $\theta = k \cdot \sigma_t$ is applied to $\mathbf{t}$: values outside the interval $[1 - k\sigma_t, 1 + k\sigma_t]$ are diminished in modulus by $k\sigma_t$ and retained, while values lying within this interval are set to the mean value $\mu_t \approx 1$ (see Figure 10.7). The resulting texture map $\mathbf{t}_\theta$, shown in Figure 10.8(b), contains spatial features that are easily detected by SAR and can be properly integrated into the MS image data by modulating the intensity of the MS image.

10.3.4.2 Generalized intensity modulation

Any linear combination of K spectral bands $M_1, M_2, \ldots, M_K$, with weights summing to unity, may be taken in principle as a generalization of intensity. If a Pan image is available to enhance spatial resolution, taking into account of the K correlation coefficients between resampled MS bands and Pan image, $\rho_{i,P}$, yields the generalized intensity (GI) as $\mathcal{I} = \sum_i w_i \cdot M_i$, in which $w_i = \rho_{i,P}/(\sum_i \rho_{i,P})$. A linear transformation $\mathbf{T}$ to be applied to the original MS pixel vectors is given by:

$$
\mathbf{T} = \begin{pmatrix}
w_1 & w_2 & w_3 & w_4 & \cdots & w_K \\
1/2 & -1/2 & 0 & 0 & \cdots & 0 \\
0 & 1/2 & -1/2 & 0 & \cdots & 0 \\
0 & 0 & 1/2 & -1/2 & \cdots & 0 \\
& & & \ddots & & \\
0 & 0 & \cdots & 0 & 1/2 & -1/2
\end{pmatrix}
\tag{10.2}
$$

If all w_is are nonzero, (10.2) is nonsingular and hence invertible. The generalized IHS transform yields the GI, $\mathcal{I}$, and $K-1$ generalized spectral differences, C_n, encapsulating the spectral information. If $\mathbf{T}$ is applied to a K-band image, with K arbitrary, and the $\mathcal{I}$ component only is manipulated, for example, sharpened by Pan details and modulated by SAR texture, the spectral information of the MS dataset is preserved, once the modulated GI, $\hat{\mathcal{I}}$, substitutes the original GI, $\mathcal{I}$, and the inverse transform $\mathbf{T}^{-1}$ is applied to obtain the fused MS image, as shown in Figure 10.6.

10.3.4.3   Simulation results

The fusion method [34] has been tested on an ERS SAR image of the city of Pavia (Italy), acquired on October 28, 2000, and on a Landsat-7/ETM+ image, with 30 m bands (1, 2, 3, 4, 5, and 7), acquired on September 22, 2000, together with the 15 m Pan image. All the optical data have been co-registered to the ERS image, which is geocoded with 12.5×12.5 m^2 pixel size. The absence of significant terrain relief would make automated registration possible, by resorting to suitable modeling of acquisition geometries of optical and SAR platforms [245, 79]. Figure 10.9 shows 512×512 details at 12.5 m pixel size. Figures 10.9(c) and 10.9(d) display true (bands 3-2-1 as R-G-B) and false (bands 5-4-3 as R-G-B) color compositions.

Figure 10.10 shows 512×512 fusion results as true and false color compositions, respectively. Urban and built-up areas, as well as roads and railway are clearly enhanced; an almost perfect preservation of spectral signatures is visible from comparisons with Figures 10.9(c) and 10.9(d).

To provide a deeper insight of the two parameters, that is, approximation depth L driving the coarseness of SAR texture and constant k tuning soft thresholding, which rules the modulation of GI by SAR texture, different fusion results are reported in Figure 10.11 for a 200×200 image fragment. Three values of L ($L = 2, 3, 4$) and four thresholds, that is, $\theta = k \cdot \sigma_t$, with $k = 0, 1, 2, 3$, have been considered ($k = 0$ indicates absence of thresholding). The most noteworthy fusion results are illustrated in Figure 10.11. For the case study of Figure 10.10, the choice $L = 3$ and $k = 1.25$ seems to provide the best trade-off between visual enhancement and accurate integration of SAR features into the MS bands. However, the choice of the parameters L and k ($1 \leq k \leq 2$ with either $L = 2$ or $L = 3$, for best results) depends on both landscape and application. How the spectral information of the original MS image is preserved is demonstrated by the values of average SAM (see Eq. (3.8)) reported in Table 10.5.

As it appears, increasing L and/or decreasing k may cause over-enhancement, which is accompanied by a loss of spectral fidelity in the fusion product. This effect is particularly evident when soft-thresholding of SAR texture is missing ($k = 0$). An analogous trend was found for the differences between the mean of original and fused bands. However, the highpass nature of the enhancing contributions from Pan and SAR makes radiometric distortion, that is, mean bias, to be negligibly small in all cases.

Supervised classification has been carried out by means of a standard maximum likelihood (ML) classifier to demonstrate the benefits of fusion. The two classes of built-up and vegetated areas, including water, have been chosen.

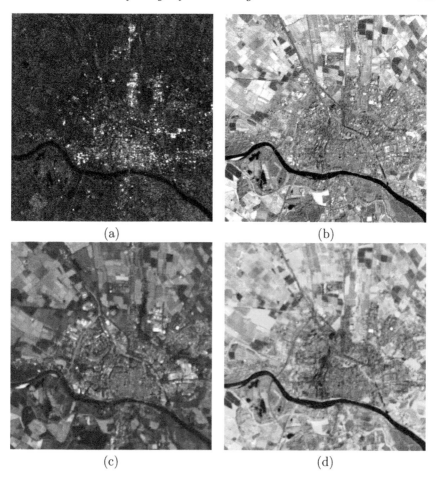

FIGURE 10.9: Original dataset of the city of Pavia (all optical data resampled at the 12.5 m scale of geocoded ERS): (a): ERS SAR (3-look amplitude); (b): Landsat-7/ETM+ Panchromatic; (c): Landsat-7/ETM+ true-color (RGB composition of 3-2-1); (d): Landsat-7/ETM+ false color (RGB composition of 5-4-3).

A detailed ground truth map was derived from a cadastral map and is portrayed in Figure 10.12(a). Buildings and man-made structures are highlighted as 17.5% of the total number of pixels. The map was manually coregistered to the available data; hence, fragmentation of classes causes misregistration of fine details, for example, buildings surrounded by vegetated areas in outskirts and small vegetation patches embedded in the city center, with large fraction of mixed pixels, thereby making classification a challenging task to be performed automatically. The classification map obtained from fused data is displayed in Figure 10.12(b) and fairly matches the extremely detailed ground

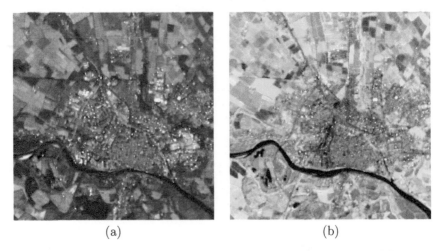

FIGURE 10.10: Results of ETM+ and ERS fusion with $L = 3$ and $k = 1.25$: (a): 3-2-1 true-color composition; (b): 5-4-3 false-color composition.

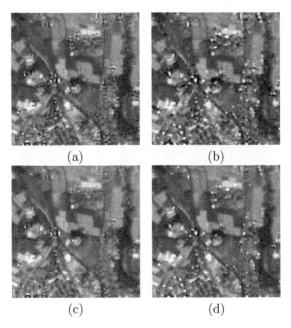

FIGURE 10.11: True-color 200×200 details of fusion products. (a): $L = 2$ and $k = 1$; (b): $L = 3$ and $k = 1$; (c): $L = 2$ and $k = 1.5$; (d): $L = 3$ and $k = 1.5$.

truth map representing a heterogeneous blend of built-up subclasses, including industrial areas.

Confusion matrices are reported in Table 10.6 for three distinct simula-

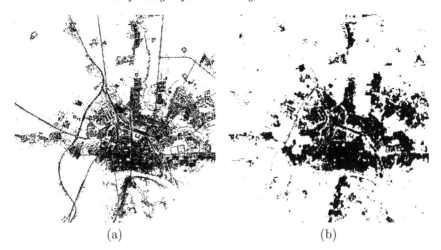

(a)   (b)

FIGURE 10.12: (a): Ground truth map detailing buildings and man-made structures; (b): map obtained from ML classification of the six MS bands fused with Pan and SAR.

TABLE 10.5: Spectral distortion (average spectral angle [SAM] between the resampled original and the fused MS pixels) vs. approximation level L, and soft threshold of the modulating texture, $\theta = k\sigma_t$.

	$k = 0$	$k = 1$	$k = 2$	$k = 3$
$L = 2$	1.102°	0.510°	0.413°	0.395°
$L = 3$	2.123°	0.672°	0.439°	0.405°
$L = 4$	2.777°	0.720°	0.449°	0.415°

tions, in which classification of the six fused bands is compared with those carried out from the eight individual images — the six MS bands plus the Pan and SAR images — and from the resampled six MS bands only. The training set was chosen on areas having either different degrees of building density, or different types of fields (including water), to better represent the intrinsic heterogeneity of both the urban and vegetation classes. The percentage of training pixels is 7.5%. The upper entry in Table 10.6 shows results when the test set is independent of the training set and its size is equal to the remaining 92.5% of pixels. When SAR data are used, the overall accuracy is obviously greater, by more than 3%, than that achieved from MS data only. However, the introduction of SAR does not improve classification of vegetation, but significantly enhances (by almost 15%) the discrimination capability of urban areas. The difference in performance between fused and *unfused* optical and SAR data is subtler: 0.2% in overall accuracy and 1.2% in classification accuracy of urban areas. The lower entry in Table 10.6 shows that classifica-

TABLE 10.6: Confusion matrices of urban (U) and vegetation (V) classes for the three cases of MS bands fused with Pan & SAR, MS+Pan+SAR separately, and MS bands only. Upper table: training set 7.5% of pixels; test set the remaining 92.5%. Lower table: training set 7.5%; test set the same 7.5%.

MS Fused w/ P&SAR		MS+P+SAR		MS Only	
Accuracy 88.7%		Accuracy 88.5%		Accuracy 85.4%	

	U	V	U	V	U	V
U	77.2%	9.9%	76.0%	9.9%	62.6%	9.8%
V	22.8%	90.1%	24.0%	90.1%	37.4%	90.2%

MS Fused w/ P&SAR		MS+P+SAR		MS Only	
Accuracy 99.5%		Accuracy 99.4%		Accuracy 99.5%	

	U	V	U	V	U	V
U	99.4%	0.4%	99.4%	0.9%	99.0%	0.6%
V	0.6%	99.6%	0.6%	99.1%	1.0%	99.6%

tion accuracy is dramatically higher when it is measured on training pixels only. The large discrepancy with the values obtained on the test set (92.5% of pixels) suggests that a certain misregistration of the ground truth map is likely to have occurred.

10.3.4.4 Summary and discussion

A two-step fusion method based on MRA achieved through ATW has proven itself to be promising for multisensor image fusion of SAR and multiband optical data. The procedure allows homogeneous data (MS and Pan) to be integrated within the physically heterogeneous SAR backscatter. The method does not directly merge spectral radiances and SAR reflectivity; instead, the combination is obtained by modulating the intensity of MS bands, after sharpening by injection of Pan highpass details, through texture extracted from SAR. An arbitrary number of spectral bands can be dealt with, thanks to the definition a generalized intensity, tailored to the spectral correlation of MS and Pan data.

Simulations carried out on Landsat-7 ETM+ and ERS data of an urban area demonstrate careful preservation of spectral signatures on vegetated regions, bare soil, and also on built-up areas (buildings and road network) where information from SAR texture enhances the fusion product both for visual analysis and classification purposes. Unfortunately, multiple resampling of the available dataset have impaired SAR texture to a large extent. Thus, visual results of fusion are less alluring than one might expect. Objective evaluations of the quality of fusion results have been presented both as spectral fidelity scores with the original MS data, and as classification accuracies obtained either with or without fusion. Thanks to the minimal spectral distortion introduced in the optical data, fusion is not penalized, which may occur sometimes, but slightly rewarding in terms of overall accuracy. The fusion procedure yields small but steady benefits for automated discrimination of manmade structures typical of urban areas.

10.4 Concluding Remarks

Problems that may arise in the fusion of data acquired by heterogeneous sensors have been addressed in this chapter. Two widespread cases and related examples involving fusion of images acquired by thermal, optical, and SAR sensors have been reported. These examples are aimed at highlighting that the straightforward use of existing solutions customized for different datasets should not be applied without a preliminary analysis of the imaged scene in order to capture the different natures of the datasets.

The sharpening of thermal bands can substantially brought back to a pansharpening without spectral overlap between two imaging instruments that count photons in both cases; thus, solutions customized for the pansharpening of the SWIR bands of an HS image may be profitably adopted. Instead, optical and SAR fusion, whose product is an optical image, and especially the reverse problem, whose solution yields a SAR image, requires the study and development of new methodologies other than pansharpening, or at least not directly linked to this.

Unlike conventional pansharpening methods that are applied between data of the same typology, for which established quality evaluation tools exist, assessing fusion results for heterogeneous datasets, especially optical and SAR, is not a trivial task. A viable strategy could be introducing application-oriented evaluations, like thematic classification carried out on the same datasets with and without a preliminary fusion. Such evaluations that are in general not advisable in the framework of MS / HS pansharpening, might become acceptable in specific heterogeneous contexts.

Chapter 11

New Trends of Remote Sensing Image Fusion

11.1 Introduction

Recent trends in image fusion, including remote sensing applications, involve the super-resolution (SR) paradigm and, more generally, apply constrained optimization algorithms to solve the ill-posed problem of spectral-spatial (pansharpening) and spatio-temporal image resolution enhancement.

Conventional approaches to generating a SR image normally require input multiple spatial/spectral/temporal low-resolution images of the same scene. The SR task is cast as the inverse problem of recovering the original high-resolution image by fusing the low-resolution images, based upon reasonable assumptions or prior knowledge about the observation model that maps the high-resolution image to the low-resolution ones. The fundamental reconstruction constraint for SR is that the recovered image, after applying the same generation model, should reproduce the observed low-resolution images. However, SR image reconstruction is generally a severely ill-posed problem because of the insufficient number of low-resolution images, ill-conditioned registration and unknown blurring operators, and the solution from the reconstruction constraint is not unique. Various regularization methods have been proposed to further stabilize the inversion of this ill-posed problem [270].

A similar approach considers image fusion as a restoration problem. The aim is therefore to reconstruct the original scene from a degraded observation, or, equivalently, to solve a classical deconvolution problem [155]. Prior knowledge is required on the nature of the two-dimensional convolution that models the band-dependent point spread function of the imaging system.

Bayesian methods and variational methods have been also proposed in the last decade, with different possible solutions which are based on specific assumptions that make the problem mathematically tractable.

Alternative pansharpening methods adopt new spectral representations of multiband images, which assume that an image can be modeled as a linear composite of a foreground component carrying the spatial detail and a background color component conveying the sensed spectral diversity. The advantage of this new approach is that it does not rely on any image formation assumption as in the case of restoration or variational or sparse decomposition approaches.

Most methods based on these new approaches suffer, on the one hand, from modeling inaccuracies, and on the other hand, on high computational complexity that limits their applicability to practical remote sensing applications.

11.2   Restoration-Based Approaches

A class of recently developed image fusion methods considers pansharpening as a restoration problem. The aim of pansharpening is therefore to reconstruct the original scene from a degraded observation, or, equivalently, to solve a classical deconvolution problem [155]. Following this approach, each band of a multispectral image, neglecting additive noise, can be modeled as the two-dimensional convolution of the corresponding band at a high-spatial resolution, with a linear shift-invariant blur, that is, the band-dependent point spread function of the imaging system.

We refer to $\tilde{M}_k$ as the original multispectral images M_k resampled to the scale of the panchromatic band P (of size $M \times N$ pixels). A degradation model is introduced, for which $\tilde{M}_k$ can be obtained as noisy blurred versions of the ideal multispectral images $\bar{M}_k$,

$$\tilde{M}_k = H_k * \bar{M}_k + v_k \quad k = 1, \ldots, N_b \tag{11.1}$$

where $*$ denotes the 2-D convolution operation, H_k is the point spread function (PSF) operator for the kth band, and v_k, $k = 1, \ldots, N_b$, is additive zero-mean random noise processes.

The high-resolution panchromatic image is modeled as a linear combination of the ideal multispectral images plus the observation noise

$$P = \sum_{k=1}^{N_b} \alpha_k \bar{M}_k + \Delta + w, \tag{11.2}$$

where Δ is an offset, $\alpha_k, k = 1, ..., N_b$, are the weights that satisfy the condition $\sum_{i=1}^{N_b} \alpha_k = 1$, and w is an additive zero-mean random noise [170].

The weights α_k can be calculated from normalized spectral response curves of the multispectral sensor [170], or by linear regression of the down-degraded panchromatic image P_d and the original multispectral bands M_k [26]. The offset Δ is approximately calculated using the degraded panchromatic image and the sensed low-resolution multispectral images through

$$\Delta = \frac{R^2}{M \cdot N} \sum_{m=1}^{M/R} \sum_{n=1}^{N/R} \left[P_d(m,n) - \sum_{k=1}^{N_b} \alpha_k M_k(m,n) \right] \qquad (11.3)$$

where R indicates the scale ratio between the original multispectral and panchromatic images.

The ideal high-resolution multispectral image can be estimated by solving a constrained optimization problem. In Li and Leung [170], the restored image is obtained by applying a regularized constrained least square (CLS) algorithm in the discrete sine transform (DST) domain to achieve sparse matrix computation. The solution is calculated row by row by applying the regularized pseudoinverse filter to the m-th row of the DST coefficients $\tilde{\mathbf{M}}_k$ and $\mathbf{P}$ of $\tilde{M}_k$ and P, respectively:

$$\hat{\mathbf{M}}(m) = \left(\mathbf{F}^T \mathbf{F} + \lambda \mathbf{I} \right)^{-1} \mathbf{F}^T \mathbf{F} \left[\mathbf{P}(m)^T, \tilde{\mathbf{M}}(m)^T \right]^T, \qquad m = 1, \dots, M \quad (11.4)$$

where $\mathbf{I}$ is the identity matrix and $\mathbf{F}$ is an $(N_b + 1)N \times (N_b + 1)N$ sparse matrix that is computed from the weights α_k in (11.2), the point spread function operators H_k in (11.1), and the DST transform matrix. Finally, λ is the regularization parameter that controls the degree of smoothness of the solution: when $\lambda \to 0$, (11.4) reduces to the unconstrained least squares solution, and when $\lambda \to \infty$, (11.4) becomes the ultrasmooth solution.

The main drawbacks of restoration-based methods are the inaccuracies of the observation models (11.1) and (11.2): the PSF operators H_k are assumed to be known but they often differ from their nominal values. Also, the optimal value of the regularization parameter λ is empirically calculated and can vary from sensor to sensor, and even on the particular scenario.

The adoption of transformed coefficients in the CLS solution (11.4) is required to obtain sparse matrices and reduce the computational complexity that is $\mathcal{O}(N^\beta M)$, with $2 < \beta < 3$. On the other hand, when working in a Fourier-related domain, for example, the DST, an intrinsically smoothed solution is obtained from (11.4) and poorly enhanced pansharpened images are often produced.

11.3   Sparse Representation

A new signal representation model has recently become very popular and has attracted the attention of researchers working in the field of image fusion as well as in several other areas. In fact, natural images satisfy a sparse model, that is, they can be seen as the linear combination of few elements of a dictionary or atoms. Sparse models are at the basis of compressed sensing theory (CST) [90], which is the representation of signals with a number of samples at a sub-Nyquist rate. In mathematical terms, the observed image is modeled as $y = Ax + w$, where A is the dictionary, x is a sparse vector, such that $||x||_0 \leq K$, with $K \ll M$, with M the dimension of x, and w is a noise term that does not satisfy a sparse model. In this context, fusion translates into finding the sparsest vectors with the constraint $||y - Ax||_2^2 < \epsilon$, where ϵ accounts for the noise variance. The problem is NP-hard, but it can be relaxed into a convex optimization one by substituting the pseudo-norm $|| \cdot ||_0$ with $|| \cdot ||_1$.

Recently, some image fusion methods based on the compressed sensing paradigm and sparse representations have appeared, either applied to pansharpening [167, 168, 279, 70], or to spatio-temporal fusion of multispectral images [137, 239, 138].

11.3.1   Sparse Image Fusion for Spatial-Spectral Fusion

The pioneering paper by Li and Yang [167] formulated the remote sensing imaging formation model as a linear transform corresponding to the measurement matrix in CST [90]. In this context, the high-resolution panchromatic and low-resolution multispectral images are referred as measurements, and the high-resolution MS images can be recovered by applying sparsity regularization.

Formally, it is assumed that any lexicographically ordered spatial patch of the observed images, namely y_{MS} and y_{PAN} can be modeled as

$$y = Mx + \nu \tag{11.5}$$

where $y = \begin{pmatrix} y_{MS} \\ y_{PAN} \end{pmatrix}$, $M = \begin{pmatrix} M_1 \\ M_2 \end{pmatrix}$, M_1 and M_2 indicate the decimation matrix and the panchromatic-model matrix, respectively, x is the unknown high-resolution MS image, and ν is an additive Gaussian noise term.

The goal of image fusion is to recover $\mathbf{x}$ from $\mathbf{y}$. If signal is compressible by a sparsity transform, the CoS theory ensures that the original signal can be accurately reconstructed from a small set of incomplete measurements. Thus, the signal recovering problem (11.5) can be formulated as a minimization problem with sparsity constraints

$$\hat{\alpha} = \arg \min ||\alpha||_0 \quad s.t. \ ||y - \Phi\alpha||_2^2 \leq \epsilon \tag{11.6}$$

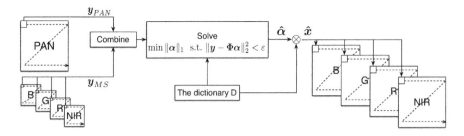

FIGURE 11.1: Flowchart of a pansharpening algorithm based on compressed sensing [167].

where $\boldsymbol{\Phi} = \boldsymbol{M}\boldsymbol{D}$, $D = (d_1, d_2, ..., d_K)$ is a dictionary, and $\boldsymbol{x} = \boldsymbol{D}\boldsymbol{\alpha}$, which explains $\boldsymbol{x}$ as a linear combination of columns from $\boldsymbol{D}$. The vector $\hat{\boldsymbol{\alpha}}$ is very sparse. Finally, the estimated $\hat{\boldsymbol{x}}$ can be obtained by $\hat{\boldsymbol{x}} = \boldsymbol{D}\hat{\boldsymbol{\alpha}}$.

The resulting pansharpening scheme is illustrated in Figure 11.1. All the patches of the panchromatic and multispectral images are processed in raster-scan order, from left-top to right-bottom with step of four pixels in the PAN image and one pixel in the MS images (1/4 ratio is assumed between PAN and MS spatial scales, as in several spaceborne sensors). First, the PAN patch y_{PAN} is combined with the MS patch y_{MS} to generate the vector $\boldsymbol{y}$. Then, the sparsity regularization (11.6) is resolved using the basis pursuit (BP) method [68] to get the sparse representation $\hat{\boldsymbol{\alpha}}$ of the fused MS image patch. Finally, the fused MS image patch is obtained by $\hat{\boldsymbol{x}} = \boldsymbol{D}\hat{\boldsymbol{\alpha}}$.

The generation of the dictionary $\boldsymbol{D}$ is the key problem of all CoS-based pansharpening approaches. In Li and Yang [167], the dictionary was generated by randomly sampling raw patches from high-resolution MS satellite images. Since such images are not available in practice, [167] reduces to a theoretical investigation on the applicability of CST to pansharpening. More recent papers have proposed different solutions to this problem, in order to deal with practical remote sensing applications. In Li, Yin, and Fang [168], the sparse coefficients of the PAN image and low-resolution MS image are obtained by the orthogonal matching pursuit algorithm. Then, the fused high-resolution MS image is calculated by combining the obtained sparse coefficients and the dictionary for the high-resolution MS image. The main assumption is that the dictionaries $\boldsymbol{D}_h^{ms}$, $\boldsymbol{D}^{pan}$ and $\boldsymbol{D}_l^{ms}$ have the relationships:

$$\boldsymbol{D}^{pan} = \boldsymbol{M}_2 \boldsymbol{D}_h^{ms} \tag{11.7}$$

$$\boldsymbol{D}_l^{ms} = \boldsymbol{M}_1 \boldsymbol{D}_h^{ms}. \tag{11.8}$$

First, $\boldsymbol{D}^{pan}$ and $\boldsymbol{D}_l^{ms}$ are computed from randomly selected samples of the available PAN and MS data by applying the K-SVD method [3]. The dictionary $\boldsymbol{D}_h^{ms}$ is estimated by applying an iterative gradient descent method to solve a minimization problem based on the MS dictionary model (11.8).

Obviously, the computational complexity of the method is huge, while the improvement with respect to effective classical pansharpening algorithms is negligible. As an example, the algorithm proposed in Li, Yin, and Fang [168], requires about 15 min on a very small (64 × 64) MS image, while, by considering the same hardware and programming software configurations, pansharpening methods based on MRA [12] or component substitution [119] provide pansharpened images with the same quality (measured by QNR, Q4, ERGAS score indexes) in one or few seconds.

Different from Li, Yin, and Fang [168], the method proposed in Zhu and Bamler [279], named Sparse Fusion of Images (SparseFI), explores the sparse representation of multispectral image patches in a dictionary trained only from the panchromatic image at hand. Also, it does not assume any spectral composition model of the panchromatic image, that is, it does not adopt a composition model similar to (11.2), which implies a relationship between the dictionaries for PAN and MS, as in (11.7). The method is described synthetically by the scheme reported in Figure 11.2.

P is a matrix that extracts the region of overlap between the current target patch and previously reconstructed ones, while w_k contains the pixel values of the previously reconstructed HR multispectral image patch on the overlap region. Parameter β is a weighting factor that gives a trade-off between goodness of fit of the LR input and the consistency of reconstructed adjacent HR patches in the overlapping area. The algorithm performances are not outstanding [279], since it provides pansharpened images with similar quality of adaptive IHS fused products.

In Cheng, Wang, and Li [70], a method is proposed to generate the high-resolution multispectral (HRM) dictionary from HRP and LRM images. The method includes two steps. The first step is AWLP pansharpening to obtain preliminary HRM images. The second step performs dictionary training using patches sampled from the results of the first step. As in Li, Yin, and Fang [168], a dictionary training scheme is designed based on the well-known K-SVD method. The training process incorporates information from the HRP image, which improves the ability of the dictionary to describe spatial details. While better quality score indexes are obtained with respect to the boosting pansharpening method AWLP, no remarkable improvements are introduced by this method with respect to fast and robust classical component substitution methods such as GSA-CA (Aiazzi *et al.* [25]), as reported in Cheng, Wang, and Li [70].

In Huang *et al.* [138], a spatial and spectral fusion model based on sparse matrix factorization is proposed and tested on Landsat 7 and MODIS acquisitions at the same date. The model combines the spatial information from sensors with high-spatial resolution, with the spectral information from sensors with high-spectral resolution. A two-stage algorithm is introduced to combine these two categories of remote sensing data. In the first stage, an optimal spectral dictionary is obtained from data with low-spatial and high-spectral resolution to represent the spectral signatures of various materials in the scene. Given the simple observation that there are probably only a few

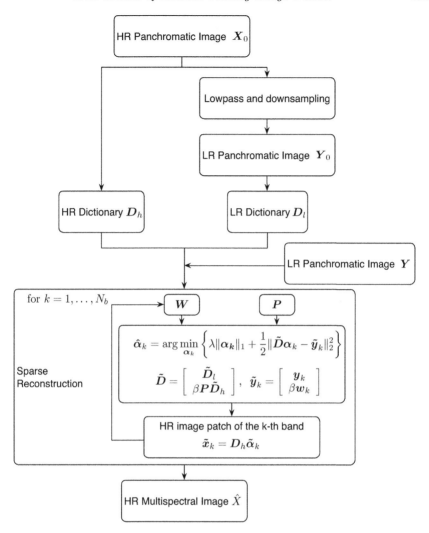

FIGURE 11.2: Block diagram of the pansharpening method proposed in Zhu and Bamler [279].

land surface materials contributing to each pixel in this kind of images, the problem is formalized as a sparse matrix factorization problem. In the second stage, by using the spectral dictionary developed in the first stage, together with data with high-spatial and low-spectral resolution, the spectrum of each pixel is reconstructed to produce a high-spatial and high-spectral resolution image via a sparse coding technique.

In synthesis, a clustering- or vector-quantization-based method is adopted to optimize a dictionary on a set of image patches by first grouping patterns such that their distance to a given atom is minimal, and then updating the atom such that the overall distance in the group of patterns is minimal. This process assumes that each image patch can be represented by a single atom in the dictionary, and this reduces the learning procedure to a K-means clustering. A generalization of this method for dictionary learning is the K-singular value decomposition (K-SVD) algorithm [3], which represents each patch by using multiple atoms with different weights. In this algorithm, the coefficient matrix and basis matrix are updated alternatively.

11.3.2 Sparse Spatio-Temporal Image Fusion

Most instruments with fine spatial resolution (e.g., SPOT and Landsat TM with a 10 m and 30 m spatial resolution) can only revisit the same location on Earth at intervals of half to one month, while other instruments with coarse spatial resolution (e.g., MODIS and SPOT VEGETATION with a 250–1000-m spatial resolution) can make repeated observations in one day. As a result, there is so far still no sensor that can provide both high spatial resolution (HSR) and frequent temporal coverage. One possible cost-effective solution is to explore data integration methods that can blend the two types of images from different sensors to generate high-resolution synthetic data in both space and time thereby enhancing the capability of remote sensing for monitoring land surface dynamics, particularly in rapidly changing areas. In the example in Figure 11.3, the goal is to predict the unavailable high-spatial-resolution Landsat image at date t_2 from the Landsat images at dates t_1 and t_3 and the low-spatial-resolution MODIS acquisitions at dates t_1, t_2, t_3.

One critical problem that should be addressed by a spatio-temporal reflectance fusion model is the detection of the temporal change of reflectance over different pixels during an observation period. In general, such change encompasses both phenology change (e.g., seasonal change of vegetation) and type change (e.g., conversion of bare soil to concrete surface), and it is considered more challenging to capture the latter than the former in a fusion model.

In Huang and Song [137], a data fusion model, called SParse-representation-based Spatio-Temporal reflectance Fusion Model (SPSTFM), is proposed, which accounts for all the reflectance changes during an observation period, whether type or phenology change, in a unified way by sparse representation. It allows for learning the structure primitives of signals via an overcomplete dictionary and reconstructing signals through sparse coding. SPSTFM learns the differences between two HSR images and their corresponding LSR acquisitions from a different instrument via sparse signal representation. It can predict the high-resolution difference image (HRDI) more accurately than searching similar neighbors for every pixel because it considers the structural similarity (SSIM), particularly for land-cover type changes. Rather than supposing a linear change of reflectance as in previous methods, sparse representation can obtain the change prediction in an intrinsic nonlinear form

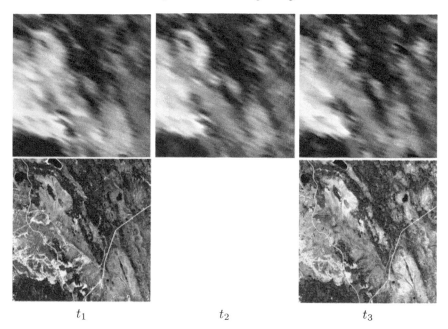

$$t_1 \qquad\qquad\qquad t_2 \qquad\qquad\qquad t_3$$

FIGURE 11.3: Predicting the Landsat image at date t_2 from Landsat images at dates t_1 and t_3 and MODIS images at all dates.

because sparse coding is a nonlinear reconstruction process through selecting the optimal combination of signal primitives.

Formally, the Landsat image and the MODIS image are denoted as L_i and M_i on t_i date, respectively, where the MODIS images are extended to have the same size as Landsat via bilinear interpolation. Let $Y_{i,j}$ and $X_{i,j}$ represent the HRDI and LRDI between t_i and t_j, respectively, and their corresponding patches are $y_{i,j}$ and $x_{i,j}$, which are formed by putting patches into column vectors. The relationship diagram for these variables is reported in Figure 11.4. L_2 can then be predicted as follows:

$$L_2 = W_1 * (L_1 + \hat{Y}_{21}) + W_3 * (L_3 - \hat{Y}_{32}) \tag{11.9}$$

In order to estimate $\hat{Y}_{21}$ and $\hat{Y}_{32}$ in (11.9), the dictionary pair D_l and D_m must be formulated. The two dictionaries D_l and D_m are trained using the HRDI and LRDI patches between t_1 and t_3, respectively, according to the following optimization

$$\{D_l^*, D_m^*, \Lambda^*\} = \arg\min_{D_l, D_m, \Lambda} \left\{ \|Y - D_l\Lambda\|_2^2 + \|X - D_m\Lambda\|_2^2 + \lambda\|\Lambda\|_1 \right\} \tag{11.10}$$

where Y and X are the column combination of lexicographically stacking image patches, sampled randomly from Y_{13} and X_{13}, respectively. Similarly,

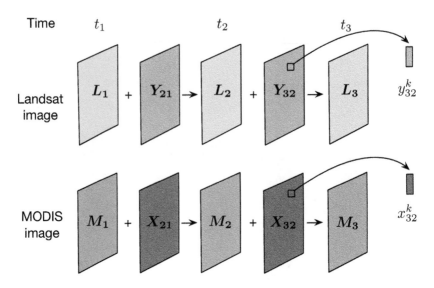

FIGURE 11.4: Block diagram of the spatio-temporal fusion proposed in Huang and Song [137].

Λ is the column combination of representation coefficients corresponding to every column in Y and X.

A different approach has been proposed in Song and Huang [239], which adopts a two-step procedure to avoid large prediction errors due to the large spatial resolution difference between MODIS and Landsat 7 data. First, it improves the spatial resolution of MODIS data and then it fuses the MODIS with an improved spatial resolution and the original Landsat data.

Denote the MODIS image, the Landsat image, and the predicted transition image on t_i as M_i, L_i, and T_i, respectively. The super-resolution of MODIS data contains two steps: the dictionary-pair training on known M_1 and L_1 and the transition image prediction. For training a dictionary pair, the high-resolution image features and low-resolution image features are extracted from difference image space of $L_1 - M_1$ and gradient feature space of M_1 in patch form (e.g., 5×5), respectively. Stacking these feature patches into columns forms the training sample matrices Y and X, where Y and X stand for high-resolution samples and low-resolution samples, respectively, and their columns are in correspondence. First, the low-resolution dictionary D_l is derived by applying the K-SVD [19] training procedure on X via optimizing the following objective function:

$$\{D_l^*, \Lambda^*\} = \underset{D_l, \Lambda}{\arg\min} \left\{ \|X - D_l\Lambda\|_F^2 \right\} \qquad \text{s.t.} \quad \forall i, \|\alpha_i\|_0 \leq K_0 \qquad (11.11)$$

where Λ is a column combination of representation coefficients corresponding to every column in X.

To establish a correspondence between high-resolution and low-resolution training samples, the high-resolution dictionary is constructed by minimizing the approximation error on Y with the same sparse representation coefficients Λ^* in (11.11), that is,

$$D_h^* = \arg\min_{D_h} \|Y - D_h\Lambda^*\|_F^2. \tag{11.12}$$

The solution of this problem can be directly derived from the following pseudoinverse expression (given that Λ^* has full row rank):

$$D_h = Y(\Lambda^*)^+ = Y\Lambda^{*T}(\Lambda^*\Lambda^{*T})^{-1}. \tag{11.13}$$

To predict the transition image T_2 from M_2, the same gradient features X_2 are extracted from M_2 as in the training process. Denote the ith column of X_2 as x_{2i}; then, its sparse coefficient α_i with respect to dictionary D_i can be obtained by employing the sparse coding technique called orthogonal matching pursuit (OMP). Because the corresponding high-resolution sample and low-resolution sample are enforced and represented by the same sparse coefficients with respect to D_h and D_l, respectively, the corresponding ith middle-resolution patch column y_{2i} can be predicted by $y_{2i} = D_h * \alpha_i$. The other middle-resolution patch columns can be predicted by this same process. After transforming all columns y_{2i} into a patch form, the difference image Y_2 between T_2 and M_2 is predicted. Thus, T_2 is reconstructed by $T_2 = Y_2 + M_2$. For the fusion procedure in the next stage, the transition image T_1 is also predicted in the same procedure. Here, the transition images T_1 and T_2 have the same size and extent as that of L_1 and L_2.

Finally, Landsat 7 and transition images are fused via highpass modulation (HPM)

$$L_2 = T_2 + \left(\frac{T_2}{T_1}\right)[L_1 - T_1]. \tag{11.14}$$

This fusion is in accordance with a linear temporal change model between T_1 and T_2.

In general, experiments show that spatio-temporal fusion based on sparse representation performs better on phenology change than type change. This can be interpreted in terms of sparsity theory, that is, more representation errors usually arise when there are more complex signals to be represented. Further work is also needed to reduce the computational complexity of spatio-temporal fusion approaches based on sparse representation.

11.4 Bayesian Approaches

In its most general formulation [276], the problem of Bayesian image fusion can be described as the fusion of an HS image ($\mathbf{Y}$) with low-spatial resolution and high-spectral resolution and an MS image ($\mathbf{X}$) with high-spatial resolution and low-spectral resolution. Ideally, the fused result $\mathbf{Z}$ has the spatial resolution of $\mathbf{X}$ and the spectral resolution of $\mathbf{Y}$. It is assumed that all images are equally spatially sampled at a grid of N pixels, which is sufficiently fine to reveal the spatial resolution of $\mathbf{X}$. The HS image has N_b spectral bands, and the MS image has N_h ($N_h < N_b$) bands, with $N_h = 1$ in the case of a panchromatic band (pansharpening case).

By denoting images column-wise lexicographically ordered for matrix notation convenience, as in the case of $\mathbf{Z} = [\mathbf{Z}_1^T, \mathbf{Z}_2^T, \ldots, \mathbf{Z}_N^T]^T$, where $\mathbf{Z}_i$ denotes the column vector representing the i-th pixel of $\mathbf{Z}$, the imaging model between $\mathbf{Z}$ and $\mathbf{Y}$ can be written as

$$\mathbf{Y} = \mathbf{WZ} + \mathbf{N}, \tag{11.15}$$

where $\mathbf{W}$ is a potentially wavelength-dependent spatially varying system point spread function (PSF) which performs blurring on $\mathbf{Z}$. $\mathbf{N}$ is modeled as multivariate Gaussian-distributed additive noise with zero mean and covariance matrix $\mathbf{C_N}$, independent of $\mathbf{X}$ and $\mathbf{Z}$. Between $\mathbf{Z}$ and $\mathbf{X}$, a jointly normal model is generally assumed.

The approach to the pansharpening problem within a Bayesian framework relies on the statistical relationships between the various spectral bands and the panchromatic band. In a Bayesian framework, an estimation of $\mathbf{Z}$ is obtained as

$$\hat{\mathbf{Z}} = \arg \max_{\mathbf{Z}} p(\mathbf{Z}|\mathbf{Y}, \mathbf{X}) = \arg \max_{\mathbf{Z}} p(\mathbf{Y}|\mathbf{Z})p(\mathbf{Z}|\mathbf{X}). \tag{11.16}$$

Generally, the first probability density function $p(\mathbf{Y}|\mathbf{Z})$ of the product in Eq. (11.16) is obtained from an observation model (11.15) where the PSF $\mathbf{W}$ reflects the spatial blurring of the observation $\mathbf{Y}$ and $\mathbf{N}$ reflects the additive Gaussian white noise with covariance matrix $\mathbf{C_N}$. The second pdf $p(\mathbf{Z}|\mathbf{X})$ in Eq. (11.16) is obtained from the assumption that $\mathbf{Z}$ and $\mathbf{X}$ are jointly normally distributed. This leads to a multivariate normal density for $p(\mathbf{Z}|\mathbf{X})$.

Different solutions have been proposed, which are based on specific assumptions that make the problem mathematically tractable.

In Fasbender, Radoux, and Bogaert [98], a simplified model is assumed first, $\mathbf{Y} = \mathbf{Z} + \mathbf{N}$, not accounting for the modulation transfer function of the imaging system, then a linear-regression model that links the multispectral pixels to the panchromatic ones is considered, and finally a noninformative prior pdf is adopted for the image $\mathbf{Z}$ to be estimated.

In Zhang, De Backer, and Scheunders [276], the estimation problem is approached in the domain of the à-trous wavelet coefficients. Since the applied à-trous transformation is a linear operation, the same model in Eq. (11.15) holds for each of the obtained detail images and the same estimation in (11.16) can be adopted for the transformed coefficients at each scale. The advantage of the application of both models in the wavelet domain is that they are applied at each orientation and resolution level, with a separate estimation of the covariances for each level. This allows for a resolution- and orientation-specific adaptation of the models to the image information, which is advantageous for the fusion process.

In Zhang, Duijster, and Scheunders [277], a Bayesian restoration approach is proposed. The restoration is based on an expectation-maximization (EM) algorithm, which applies a deblurring step and a denoising step iteratively. The Bayesian framework allows for the inclusion of spatial information from the high-spatial resolution image (multispectral or panchromatic) and accounts for the joint statistics with the low-spatial resolution image (possibly a hyperspectral image).

The key concept in the EM-based restoration procedure is that the observation model in Eq. (11.15) is inverted by performing the deblurring and denoising in two separate steps. To accomplish this, the observation model is decomposed as

$$\mathbf{Y} = \mathbf{W}\mathbf{X} + \mathbf{N}'' \qquad (11.17)$$

$$\mathbf{X} = \mathbf{Z} + \mathbf{N}' \qquad (11.18)$$

In this way, the noise is decomposed into two independent parts $\mathbf{N}'$ and $\mathbf{N}''$, with $\mathbf{W}\mathbf{N}' + \mathbf{N}'' = \mathbf{N}$.

Choosing $\mathbf{N}'$ to be white facilitates the denoising problem (11.18). However, $\mathbf{W}$ colors the noise so that $\mathbf{N}''$ becomes colored.

Equations (11.17) and (11.18) are iteratively solved using the EM algorithm. An estimation of $\mathbf{Z}$ is obtained from a restoration of the observation $\mathbf{Y}$ combined with a fusion with the observation $\mathbf{X}$.

Bayesian approaches to pansharpening suffer from modeling errors due to simplifications that are intentionally introduced to reduce the computational complexity as in Fasbender, Radoux, and Bogaert [98] where the MTFs of the imaging sensors are not considered.

Also, iterative processing and numerical instability make Bayesian approaches more complex and less reliable for practical remote sensing image fusion applications on true image data than multiresolution-based or component substitution fusion algorithms.

11.5   Variational Approaches

Pansharpening is in general an ill-posed problem that needs regularization for optimal results. The approach proposed in Palsson, Sveinsson, and Ulfarsson [203] uses total variation (TV) regularization to obtain a solution that is essentially free of noise while preserving the fine detail of the PAN image. The algorithm uses majorization-minimization (MM) techniques to obtain the solution in an iterative manner.

Formally, the dataset consists of a high-spatial resolution panchromatic image $\boldsymbol{y}_{PAN}$ and the low-spatial resolution multispectral image $\boldsymbol{y}_{MS}$. The PAN image has dimensions four times larger than the MS image thus the ratio in pixels is 1 to 16. The MS image contains 4 bands, RGB and near-infrared (NIR). The PAN image is of dimension $M \times N$ and the MS image is of dimension $m \times n$, where $m = M/4$ and $n = N/4$.

There are two assumptions that define the model. The first is that the low-spatial resolution MS image can be described as a degradation (decimation) of the pansharpened image $\boldsymbol{x}$. In matrix notation, $\boldsymbol{y}_{MS} = \boldsymbol{M}_1 \boldsymbol{x} + \boldsymbol{\epsilon}$ where

$$\boldsymbol{M}_1 = \frac{1}{16} \mathbf{I}_4 \otimes ((\mathbf{I}_n \otimes \mathbf{1}_{4\times1}^T) \otimes (\mathbf{I}_m \otimes \mathbf{1}_{4\times1}^T)) \tag{11.19}$$

is a decimation matrix of size $4mn \times 4MN$, $\mathbf{I}_4$ is an identity matrix of size 4×4, $\otimes$ is the Kronecker product, and $\boldsymbol{\epsilon}$ is zero-mean Gaussian noise.

The second assumption is that the PAN image is a linear combination of the bands of the pansharpened image with some additive Gaussian noise. This can be written in the matrix notation as $\boldsymbol{y}_{PAN} = \boldsymbol{M}_2 \boldsymbol{x} + \boldsymbol{\epsilon}$, where $\boldsymbol{\epsilon}$ is zero-mean Gaussian noise and

$$\boldsymbol{M}_2 = [\omega_1 \mathbf{I}_{MN}, \omega_2 \mathbf{I}_{MN}, \omega_3 \mathbf{I}_{MN}, \omega_4 \mathbf{I}_{MN}] \tag{11.20}$$

where $\omega_1, \ldots, \omega_4$ are constants that sum to one. These constants determine the weight of each band in the PAN image.

Now $\boldsymbol{M}_1$ and $\boldsymbol{M}_2$ have the same number of columns and thus the expressions for $\boldsymbol{y}_{MS}$ and $\boldsymbol{y}_{PAN}$ can be combined into a single equation, producing the classical observational model

$$\boldsymbol{y} = \boldsymbol{M} \boldsymbol{x} + \boldsymbol{\epsilon} \tag{11.21}$$

where $\boldsymbol{y} = [\boldsymbol{y}_{MS}^T \boldsymbol{y}_{PAN}^T]^T$ and $\boldsymbol{M} = [\boldsymbol{M}_1^T \boldsymbol{M}_2^T]^T$.

One can define the TV of the MS image as

$$\mathrm{TV}(\boldsymbol{x}) = \left\| \sqrt{(\boldsymbol{D}_H \boldsymbol{x})^2 + (\boldsymbol{D}_V \boldsymbol{x})^2} \right\|_1 \tag{11.22}$$

where $\boldsymbol{x}$ is the vectorized 4-band MS image, $D_H = (\boldsymbol{I}_4 \otimes \boldsymbol{D}_H)$, $D_V = (\boldsymbol{I}_4 \otimes \boldsymbol{D}_V)$ and the matrices $\boldsymbol{D}_H$ and $\boldsymbol{D}_V$ are defined such that when multiplied by a vectorized image they give the first-order differences in the horizontal direction and vertical direction, respectively. The cost function of the TV regularized problem can be formulated as

$$J(\boldsymbol{x}) = \|\boldsymbol{y} - \boldsymbol{Mx}\|_2^2 + \lambda \, \mathrm{TV}(\boldsymbol{x}). \tag{11.23}$$

Minimizing this cost function is difficult because the TV functional is not differentiable. However, MM techniques can be used to replace this difficult problem with a sequence of easier ones

$$\boldsymbol{x}_{k+1} = \arg\min_{\boldsymbol{x}} \ Q(\boldsymbol{x}, \boldsymbol{x}_k) \tag{11.24}$$

where $\boldsymbol{x}_k$ is the current iterate and $Q(\boldsymbol{x}, \boldsymbol{x}_k)$ is a function that maximizes the cost function $J(\boldsymbol{x})$. This means that $Q(\boldsymbol{x}, \boldsymbol{x}_k) \geq J(\boldsymbol{x})$ for $\boldsymbol{x} \neq \boldsymbol{x}_k$ and $Q(\boldsymbol{x}, \boldsymbol{x}_k) = J(\boldsymbol{x})$ for $\boldsymbol{x} = \boldsymbol{x}_k$. By iteratively solving (11.24), $\boldsymbol{x}_k$ will converge to the global minimum of $J(\boldsymbol{x})$.

A majorizer for the TV term can be written using the matrix notation as

$$Q_{\mathrm{TV}}(\boldsymbol{x}, \boldsymbol{x}_k) = \boldsymbol{x}^T \boldsymbol{D}^T \boldsymbol{\Lambda}_k \boldsymbol{D} \boldsymbol{x} + c \tag{11.25}$$

where

$$\boldsymbol{\Lambda}_k = \mathrm{diag}(\boldsymbol{w}_k, \boldsymbol{w}_k) \ \ \text{with} \ \ \boldsymbol{w}_k = \left(2\sqrt{(\boldsymbol{D}_H \boldsymbol{x}_k)^2 + (\boldsymbol{D}_V \boldsymbol{x}_k)^2}\right)^{-1} \tag{11.26}$$

and the matrix $\boldsymbol{D}$ is defined as $\boldsymbol{D} = \left[\boldsymbol{D}_H^T \boldsymbol{D}_V^T\right]^T$.

By defining

$$Q_{\mathrm{DF}}(\boldsymbol{x}, \boldsymbol{x}_k) = (\boldsymbol{x} - \boldsymbol{x}_k)^T (\alpha \boldsymbol{I} - \boldsymbol{M}^T \boldsymbol{M})(\boldsymbol{x} - \boldsymbol{x}_k), \tag{11.27}$$

the function to minimize becomes

$$Q(\boldsymbol{x}, \boldsymbol{x}_k) = \|\boldsymbol{y} - \boldsymbol{Mx}\|_2^2 + Q_{\mathrm{DF}}(\boldsymbol{x}, \boldsymbol{x}_k) + \lambda Q_{\mathrm{TV}}(\boldsymbol{x}, \boldsymbol{x}_k). \tag{11.28}$$

It should be noted that all the matrix multiplications involving the operators $\boldsymbol{D}$, $\boldsymbol{D}^T$, $\boldsymbol{M}$, and $\boldsymbol{M}^T$ can be implemented as simple operations on multispectral images. However, the multiplication with $\boldsymbol{M}^T$ corresponds to nearest neighbor interpolation of an MS image, which is required by the problem formulation, but it provides inferior results with respect to bilinear interpolation, both according to quality metrics and visual inspection.

In general, variational methods are very sensitive to the unavoidable inaccuracies of the adopted observational model. The experimental results on true spaceborne multispectral and panchromatic images show the limitations of this class of pansharpening methods. As an example, the algorithm [203] described in this section provides fused images from QuickBird data characterized by spectral and spatial distortions [32], which are slightly lower than those obtained by a very simple (and low-performance) multiresolution-based pansharpening method, that is, a trivial coefficient substitution method in the undecimated wavelet transform (UDWT) domain: $D_\lambda = 0.042$ and $D_S = 0.027$ for [203] instead of $D_\lambda = 0.048$ and $D_S = 0.055$ for the UDWT method.

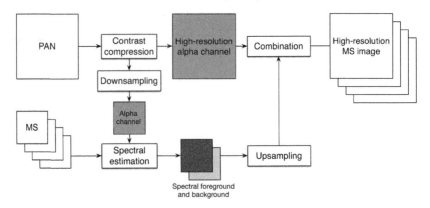

FIGURE 11.5: Block diagram of the pansharpening method proposed in Kang, Li, and Benediktsson [145].

11.6 Methods Based on New Spectral Transformations

A very recent pansharpening method [145] adopts a matting model which assumes that an image I can be modeled as a linear composite of a foreground color F and a background color B (m refers to the m-th pixel):

$$I_m = \alpha_m F_m + (1 - \alpha_m)B_m. \tag{11.29}$$

First, the spectral foreground and background of the MS image are estimated with a local linear assumption. Then, the alpha channel of the MS image is substituted by the PAN image, and a composition operation is used to reconstruct the high-resolution MS image.

According to Figure 11.5, the pansharpening method consists of five major steps: contrast compression, downsampling, spectral estimation, upsampling, and composition.

Contrast compression aims at compressing the pixel values of the panchromatic image into a relatively small range, which ensures that the edge features of the PAN image are to be processed equally in the following spectral estimation process. Two assumptions are required for the spectral estimation phase. One is that the alpha channel, spectral foreground, and background should be able to reconstruct the low-resolution MS image according to (11.29). The other is that the spectral foreground and background should be spatially smooth. The two assumptions are satisfied by solving the following energy function which has closed-form solutions:

$$\min \sum_{m \in LM} \sum_{c} \left(\widetilde{LP}_m LF_m^c + (1 - \widetilde{LP}_m) LB_m^c \right)^2$$
$$+ |\widetilde{LP}_{mx}| + \left((LF_{mx}^c)^2 + (LB_{mx}^c)^2 \right)$$
$$+ |\widetilde{LP}_{my}| + \left((LF_{my}^c)^2 + (LB_{my}^c)^2 \right) \qquad (11.30)$$

where m and c mean the m-th pixel and the c-th band and LF_{mx}^c, LF_{my}^c, LB_{mx}^c, LB_{my}^c, $\widetilde{LP}_{mx}$, and $\widetilde{LP}_{my}$ are the horizontal and vertical derivatives of the spectral foreground LF^c, spectral background LB^c, and the low-resolution alpha channel $\widetilde{LP}$.

Finally, after upsampling, LF^c, and LB^c, the pansharpened MS image is obtained by combining the high-resolution spectral foreground HF, background HB, and the high-resolution alpha channel $\widetilde{HP}$ together,

$$HM_m^c = \widetilde{HP}_m HF_m^c + (1 - \widetilde{HP}_m) HB_m^c.$$

An interesting characteristic of this method, which provides good quality pan-sharpened images, is that it does not rely on any image formation assumption which has been used by most pansharpening methods based on restoration or variational or sparse decomposition approaches. Thus, it can be applied directly for images captured by different satellites, or images that contain different types of landscape objects.

11.7   Concluding Remarks

Super-resolution, compressed sensing, Bayesian estimation, and variational theory are methodologies that have been recently applied to spectral-spatial and spatio-temporal image resolution enhancement for remote sensing applications. Specific assumptions on the image formation process and model simplifications to make the problem mathematically tractable are normally required to solve the ill-posed problems that are usually encountered through constrained optimization algorithms.

When prior knowledge about the observation model is not sufficiently verified on true image data, due to uncertainty on the band-dependent point spread function of the imaging system, or when the image reconstruction constraint is mathematically convenient but not physically consistent for the current remote sensing systems, the quality of the fusion products may decrease significantly.

Another drawback of these new approaches to remote sensing image fusion is their extremely high computational complexity. In most cases, a negligible increase in the quality of the fusion products is attained with respect to standard state-of-the-art methods at the cost of a significant increment — even by a factor of 100 or more — of the computing time. As a matter of fact, these methods are currently far from being competitive with classical approaches based on multiresolution analysis or component substitution for operational, large-scale spatial/spectral/temporal enhancement of remote sensing image data.

Chapter 12

Conclusions and Perspectives

12.1 Introduction

In the concluding section we wish to summarize the main content of this volume and highlight the key points of each topic. The following list of items constitutes the start points:

- Fusion methods have been motivated by progress in sensor technologies. Early methods represent a first generation and are unconstrained from objective quality measurements.

- Definitions of quality of fusion products are highly crucial and should rely on widely agreed concepts and on consistency measurements.

- The majority of fusion methods, including hybrid methods, can be accommodated in a unique framework that may be parametrically optimized, and constitutes the second generation of fusion methods.

- Advantages and limitations of pansharpening methods based on either CS or MRA have been investigated. The former are spatially accurate but may be spectrally inaccurate and are tolerant to aliasing of MS and moderate spatial misalignments between MS and Pan. The latter are spectrally accurate but may be spatially inaccurate and are tolerant to temporal misalignments between MS and Pan.

- An established theory and practice of pansharpening, together with

265

availability of data from upcoming missions, shall open wider horizons to HS pansharpening, thermal and optical fusion, and especially to SAR and optical fusion.

- New paradigms of signal/image processing, like compressed sensing, are largely investigated nowadays and it is expected that in some years they will provide a third generation of methods, more powerful than second-generation methods but computationally massive and presently not affordable.

In the remainder of this chapter, the individual items will be expanded into specific sections, not necessarily with the same one-to-one correspondence. The goal is to both give a synoptic presentation of materials and methods encountered throughout the volume and present a personal viewpoint of the authors on specific topics.

12.2 State of the Art of Pansharpening

The oldest and most relevant type of fusion for RS optical data is pansharpening. Beyond the plethora of different classifications and categorizations of methods over the last 25 years, in which substantially identical classes were treated as different, thereby generating a spread of classes, the great majority of pansharpening methods can be accommodated in two classes, depending on whether the detail to be injected in the MS bands interpolated at the scale of Pan is extracted by means of a spatially dispersive transformation, that is, a digital LTI filter, or not. In the latter case, the transformation yielding the details is not spatial but spectral; hence the affinity of such methods with the established class of CS methods. In the former case, extracting the high-pass component in the spatial frequencies of the Pan image is equivalent to performing an MRA of Pan.

12.2.1 Interpolation

Interpolation of the MS bands to match the spatial scale of Pan is a problem that deserves some attention. Chapter 4 is entirely devoted to this topic. If we assume that the datasets have already been coregistered, each at its own scale, interpolation must take into account the acquisition geometry of the MS and Pan instruments. In most cases concerning VHR data products, each MS pixel spans $r \times r$ Pan pixels, where r is the scale ratio between MS and Pan and is centered in the middle of the underlying $r \times r$ Pan pixel pattern. This means that if interpolation consists of replicating MS pixels r times, along both rows and columns (zero order interpolation), the resulting

MS image is exactly overlapped with the Pan image. As a consequence, interpolation of MS must use filters with linear nonzero phase to compensate the shifts intrinsically present in the original representation. Besides even-order interpolation filters (zero order or *nearest neighbor*, second order or quadratic, fourth order, and so on) also odd-order polynomial filters, like the widespread (bi)cubic interpolator, which represents the best trade-off between accuracy and computational complexity, can be designed with even length and hence linear nonzero phase [27].

The topic of interpolation with filters of odd or even lengths opens new perspectives to fusion methods based on MRA. In fact, if one uses DWT or UDWT achieved with filters of even lengths, the intrinsic shifts between MS and Pan data products is automatically compensated by the wavelet analysis and no interpolation is necessary. Conversely, ATW and GLP, employing odd-length zero phase filters, could be designed with even-length linear-phase filters in order to avoid interpolation of the original MS datasets before being incorporated as baseband of the Pan decomposition. In both cases (DWT/UDWT and ATW/GLP), computational costs are more than halved, because only Pan will be decomposed and interpolation of MS will no longer be necessary, because, for example, the top of the GLP of Pan is replaced with the MS image and the pansharpened MS is obtained by inverting the resulting pyramid.

12.2.2 Pansharpening Based on CS

CS-based pansharpening is based on some simple and fast techniques that feature spectral transformations of the original bands in new vector spaces. The rationale of CS-based fusion is that one of the transformed components (usually the first component or intensity, I) is substituted by the high-resolution panchromatic image, P, which is histogram-matched to I, before the inverse transformation is applied. However, since the histogram-matched P and I may not have the same local radiometry, spectral distortion, appearing as local color changes in a composition of three bands at a time, may occur in sharpened products. The spectral distortion can be mitigated if I is taken as a linear combination of the MS bands with weighting coefficients adjusted to the extent of the overlaps between the spectral responses of each MS channel and that of the P [251, 26]. Unlike MRA-based techniques, CS-based methods usually have a low sensitivity to spatial distortions.

12.2.3 Pansharpening Based on MRA

MRA-based techniques substantially split the spatial information of the MS bands and of the Pan image P into a series of bandpass spatial frequency channels. The high-frequency channels are inserted into the corresponding channels of the interpolated MS bands. The sharpened MS bands are synthesized from their new sets of spatial frequency channels. The ATW transform and the GLP are most widely used to perform the MRA [196, 12]. For both the methods, the zero-mean high-frequency spatial details are simply given as the difference between P and its lowpass-filtered version, P_L. The difference is that the GLP also features a decimation step with a subsequent interpolation. The decimation step has to be preceded by lowpass filtering to avoid aliasing. Recent studies [14] have demonstrated that considerable improvements in performance are achieved if the adopted lowpass filter is designed so that its frequency response matches the MTF of the spectral channel into which the details will be injected. MRA-based methods usually yield low-spectral distortions, but are very sensitive to the various typologies of spatial impairments.

12.2.4   Hybrid Methods

According to the categorization of pansharpening used in this book and originally introduced in Baronti *et al.* [45], all hybrid pansharpening schemes, which employ both a spectral transform, as CS does, and digital filter(s), as MRA does, can be brought back to either of the main classes, but in most cases to MRA. In fact, the true discrimination between class lies in the definition of the detail δ. If $\delta = P - I$, which means that the definition of detail does not imply a digital filtering of the Pan image, but only a spectral transformation, or more exactly a combination of bands to yield I, the method is of the CS type. Conversely, if $\delta = P - P_L$, where P_L is a lowpass-filtered version of Pan, the method is of the MRA type. As an example, the widely popular and top performing AWLP method [200] is of an MRA type because $\delta = P - P_L$, notwithstanding AWLP is equivalent to applying ATW to the intensity component of the nonlinear IHS transformation (6.18).

12.3   Multispectral to Hyperspectral Pansharpening

The inherent trade-off between spatial and spectral resolution has fostered in the development of remote sensing systems that include low-resolution HS coupled with high-resolution panchromatic and/or MS imaging subsystems. This provides the opportunity to jointly process the HS and higher-resolution panchromatic images to achieve high-spatial/spectral resolution images.

An accurate geometric registration of the images to be fused may be crucial, since the HS and panchromatic instrument, which are technologically and functionally separate, generally share the same optics and have a common focal plane. Thus, their focal axes cannot be parallel and coregistration of the two images cannot be a simple affine transformation of coordinates. Coregistration is particularly crucial for instruments carried on ultralight aircraft. Although radiometric calibration problems of the two instruments may exist, all pansharpening methods contain a more or less implicit phase, histogram matching for CS methods, panchromatic equalization for MRA methods, aimed at cross-calibrating Pan over MS. Analogously, atmospheric corrections of HS bands are not a prerequisite for fusion, but may be useful if pansharpening methods based on radiative transfer model are pursued. For MS pansharpening, such requirements are even less tight, because atmospheric correction is substantially the decrement by an offset of each band [174, 143]. Most notable is the offset of the blue band, caused by wavelength-selective scattering of the atmosphere that appears as at-sensor radiance without being reflected by the Earth's surface.

The main problem that occurs with HS pansharpening is that the spectral coverage of the panchromatic image does not match the wavelength acquisition range of the HS bands. As a consequence, most of the classical methods designed for MS pansharpening produce unsatisfactory results when applied to HS images since they may introduce large radiometric and spectral errors, unacceptable for both visual and quantitative analysis. Especially, CS-based methods, which rely on the spectral matching between I and P are penalized. We highlight that a CS method performs to the extent by which an I component is found from the MS bands that matches the lowpass version of Pan, P_L. Conversely, MRA methods do not rely on spectral matching, but on the correlation of individual narrow-band spectral channels and Pan. Such a correlation should be reflected by injection gains, which become crucial in this case.

To design an effective method for spatial enhancement of HS images through a panchromatic observation, exactly preserving the original spectral information of the HS data should be the main objective. To this aim, specific constraints have to be incorporated in the fusion algorithm, as in spectral angle-constrained schemes. This can be achieved by adopting suitable spatial injection models, as in modulation-based pansharpening algorithms. At the same time, limiting the spatial distortion effects due to modulation-based pansharpening will assume greater importance for achieving high-quality spatially enhanced HS images.

12.4   Fusion of Heterogeneous Datasets

As discussed in Chapter 10, pansharpening methods described in Chapter 7 cannot be straightforwardly applied to the fusion of heterogeneous datasets, because of the different nature of data to be fused. Optical and SAR image fusion is particularly crucial, for the further reason that a reference target of fusion *does not exist* instead it *exists in principle but is unavailable in practice*. A proper choice and adaptation of the most suitable method must be preliminarily envisaged in order to achieve fusion products that retain their significance when they are considered from a physical and/or informational perspective. Also, questions related to quality evaluation of fusion methods should be taken into account, as pointed out in Chapter 3 and adapted to the specific cases.

It is the opinion of the authors that the dual problem of optical-SAR fusion, that is, achieving a SAR image spectrally and/or radiometrically enhanced by VHR optical observation, will capture the attention of scientists in the coming years for the following three fundamental reasons:

1. SAR data are available under any meteorological event; optical data are not; optical images cannot be taken nighttime, SAR images can. Thus, an optically enhanced SAR image is a more reliable fusion product than a SAR-enhanced optical image.

2. There is plenty of VHR data with submetric resolution Pan that may be coupled with metric resolution second-generation SAR data products.

3. This aspect of optical-SAR fusion has practically never been investigated in the literature, at least so far. The topic is challenging, but setting some objectives requires a preliminary agreement between the researchers and users' communities, because the fusion product would be in principle general purpose and only experts could state whether it is useful for solving specific application oriented problems or not.

Unlike conventional pansharpening methods that merge datasets of the same typology, for which established quality evaluation protocols exist, assessing fusion products for heterogeneous datasets, especially optical and SAR, may be a nontrivial task. A viable strategy could be introducing application-oriented evaluations, like thematic classification carried out on the same datasets with and without a preliminary fusion. Such evaluations that are in general not advisable in the framework of MS / HS pansharpening, might become acceptable in specific heterogeneous contexts.

12.5   New Paradigms of Fusion

New trends in image fusion for remote sensing applications usually concern constrained optimization algorithms to solve the ill-posed problems of pansharpening or spatio-temporal image resolution enhancement. Super-resolution, compressed sensing, and variational theory are different methodologies that can be adopted to this aim.

While these new paradigms of fusion appear promising, at least in some cases, it seems that times are not yet mature for an extensive use of these methods for fusion of remote sensing data on operational systems providing continuity and reliability.

Specific assumptions on the image formation process and model simplifications to make the problem mathematically tractable should be required to solve the constrained optimization problems that characterize such methods. Prior knowledge about the observation model is not always verified on true image data, and often the constraints are mathematically convenient but not physically consistent.

As a matter of fact, the main drawback of these new approaches to remote sensing image fusion is their extremely high computational complexity. While the average improvement of the quality score indices reduces to less of few percents, in the most fortunate cases, the computing time for providing a sharpened MS image patch is increased by two or more orders of magnitude: the advantage over optimized second-generation fusion methods is far from being demonstrated at present time.

The methods based on compressed sensing require the estimation of dictionaries for accurate data representation in a sparse domain. Also in this case, specific assumptions that make the problem mathematically tractable are needed and random selection of image patches is performed. The physical congruence of these assumptions and the reproducibility of tests are hard to be demonstrated on true data. Sparsity theory itself indicates that more representation errors arise when there are more complex signals to be represented. This problem results in additional variability of the fusion quality depending on the data characteristics and the observed scenario.

Pansharpening approaches based on constrained optimization usually suffer from modeling errors due to simplifications that are intentionally introduced to reduce the computational complexity. Also, iterative processing and numerical instability make these approaches more complex and less reliable for practical remote sensing image fusion applications on true image data than spatially optimized multiresolution or spectrally optimized CS fusion algorithms.

12.6 Concluding Remarks

In the concluding section of the concluding chapter of this book on RS image fusion, one would expect a projection on the future of image fusion and of remote sensing in general. Instead, a simple reflection will be found. Any time, any place in the civilian world, when the gap between theory and technology has been filled, the progress could reach all layers of a society. Fifty centuries ago, the Egyptian architect built a pyramid or raised an obelisk by putting in practice, unfortunately at the cost of the lives of thousands of slaves, a personal knowledge he had developed mainly from his own skills. Apart from a few persons over the centuries, just think of Leonardo and Michelangelo, in which the gap between theory and practice was missing and a unique individual developed a theory and translated it to a technology to make artifacts, those persons who develop a theory are not the same individuals who put it in practice. Before having his head cut on the guillotine during the French Revolution, an outstanding mathematician, as Laplace was, would never conceive that the theory and transform he developed would become the basis for the control of systems after more than one century.

Over the last century, Shannon published his mathematical theory of communications in 1948. As one of the most notable achievements of such a theory, almost 40 years later the first video coders were available to users, mainly for low-bitrate videoconferencing. Approximately at the same time, the first pansharpening methods were inspired by the luminance-chrominance representation of video signals to attain exactly the opposite goal. In video coding the chrominance components are subsampled with respect to the luminance component by exploiting imperfections of the human visual system. In pansharpening, subsampled spectral components (MS) are merged with a high-resolution luminance-like component (Pan) to yield a product that is perceptually indistinguishable from the product that is achieved from nonsubsampled MS components.

Nowadays, the military technology of high-resolution imaging from space has fostered Earth observation from space. Applications so far unconceivable are now feasible. The gap between theory and technology was filled by military instruments. The abatement of costs has made traditional military technologies, like thermal and HS imaging, to also be exploited for civilian tasks. Intrinsic limitations of each imaging technology (optical, thermal, microwave, etc.) have suggested that an extra value can be reached by means of fusion. Mathematical theories, rediscovered during the last decades for signal / image processing and computer science, have fostered the development of remote sensing image analysis and fusion.

An impressive number of missions are planned or already scheduled in the coming years: Sentinel-1, RadarSat-3, COSMO-SkyMed-II, EnMap, PRISMA, Sentinel-2, GeoEye-2, MeteoSat-3, just to mention a few among those carrying imaging instruments. The technological gap is not between theory and technology, but between technology and its exploitation for any purposes that are related to the improvement in the quality of life.

Although multimedia signal and image processing is substantially disjointed from remote sensing signal and image processing / fusion, both benefit from the fact that the gap between mathematical theories and processing tools was filled during the first decades of the computer era. The theory of MRA, concisely reviewed in Chapter 5 is a notable and fitting example of this. The recent theory of compressed sensing [90] has raised the attention of both the multimedia and the Earth observation worlds, thanks to its potential for the design of new instruments and data processing tools with extra values.

It is difficult to say where this trend will lead our society. Not only in remote sensing and in image fusion there is the need for a standpoint, an equilibrium that balances the frantic rush of *innovation technology* with the social and economic benefits stemming from a well-aware exploitation of new technologies for a better life. We warmly wish that our present contribution may constitute not only a guideline in the jungle of data fusion, but also one brick of a present and upcoming sustainable development.

Bibliography

[1] N. Agam, W. P. Kustas, M. C. Anderson, Fuqin Li, and C. M. U. Neale. A vegetation index based technique for spatial sharpening of thermal imagery. *Remote Sensing of Environment*, 107(4):545–558, April 2007.

[2] M. A. Aguilar, F. Aguera, F. J. Aguilar, and F. Carvajal. Geometric accuracy assessment of the orthorectification process from very high resolution satellite imagery for Common Agricultural Policy purposes. *International Journal of Remote Sensing*, 29(24):7181–7197, December 2008.

[3] M. Aharon, M. Elad, and A. Bruckstein. K-SVD: An algorithm for designing overcomplete dictionaries for sparse representation. *IEEE Transactions on Signal Processing*, 54(11):4311–4322, November 2006.

[4] B. Aiazzi, L. Alparone, F. Argenti, and S. Baronti. Wavelet and pyramid techniques for multisensor data fusion: a performance comparison varying with scale ratios. In S. B. Serpico, editor, *Image and Signal Processing for Remote Sensing V*, volume 3871 of *Proceedings of SPIE, EUROPTO Series*, pages 251–262, 1999.

[5] B. Aiazzi, L. Alparone, F. Argenti, S. Baronti, and I. Pippi. Multisensor image fusion by frequency spectrum substitution: subband and multirate approaches for a 3:5 scale ratio case. In *Proceedings IEEE International Geoscience And Remote Sensing Symposium (IGARSS)*, pages 2629–2631, 2000.

[6] B. Aiazzi, L. Alparone, A. Barducci, S. Baronti, and I. Pippi. Multispectral fusion of multisensor image data by the generalized Laplacian pyramid. In *Proceedings of IEEE International Geoscience And Remote Sensing Symposium (IGARSS)*, pages 1183–1185, 1999.

[7] B. Aiazzi, L. Alparone, S. Baronti, V. Cappellini, R. Carlà, and L. Mortelli. A Laplacian pyramid with rational scale factor for multisensor image data fusion. In *Proceedings of International Conference on Sampling Theory and Applications - SampTA '97*, pages 55–60, 1997.

[8] B. Aiazzi, L. Alparone, S. Baronti, V. Cappellini, R. Carlà, and L. Mortelli. Pyramid-based multisensor image data fusion with enhancement of textural features. In A. Del Bimbo, editor, *Image Analysis and*

Processing, volume 1310 of *Lecture Notes in Computer Science*, pages 87–94. Springer, Berlin-Heidelberg, Germany, 1997.

[9] B. Aiazzi, L. Alparone, S. Baronti, and R. Carlà. A pyramid approach to fusion of Landsat TM and SPOT-PAN data to yield multispectral high-resolution images for environmental archaeology. In *Remote Sensing for Geography, Geology, Land Planning, and Cultural Heritage*, volume 2960 of *Proceedings of SPIE, EUROPTO Series*, pages 153–162, 1996.

[10] B. Aiazzi, L. Alparone, S. Baronti, R. Carlà, A. Garzelli, L. Santurri, and M. Selva. Effects of multitemporal scene changes on pansharpening fusion. In *Proceedings of MultiTemp 2011 - 6th IEEE International Workshop on the Analysis of Multi-temporal Remote Sensing Images*, pages 73–76, 2011.

[11] B. Aiazzi, L. Alparone, S. Baronti, R. Carlà, and L. Mortelli. Pyramid-based multisensor image data fusion. In M. A. Unser, A. Aldroubi, and A. F. Laine, editors, *Wavelet Applications in Signal and Image Processing V*, volume 3169, pages 224–235, 1997.

[12] B. Aiazzi, L. Alparone, S. Baronti, and A. Garzelli. Context-driven fusion of high spatial and spectral resolution images based on oversampled multiresolution analysis. *IEEE Transactions on Geoscience and Remote Sensing*, 40(10):2300–2312, October 2002.

[13] B. Aiazzi, L. Alparone, S. Baronti, A. Garzelli, and M. Selva. An MTF-based spectral distortion minimizing model for pan-sharpening of very high resolution multispectral images of urban areas. In *Proceedings of URBAN 2003: 2nd GRSS/ISPRS Joint Workshop on Remote Sensing and Data Fusion over Urban Areas*, pages 90–94, 2003.

[14] B. Aiazzi, L. Alparone, S. Baronti, A. Garzelli, and M. Selva. MTF-tailored multiscale fusion of high-resolution MS and Pan imagery. *Photogrammetric Engineering and Remote Sensing*, 72(5):591–596, May 2006.

[15] B. Aiazzi, L. Alparone, S. Baronti, A. Garzelli, and M. Selva. Advantages of Laplacian pyramids over "à trous" wavelet transforms. In L. Bruzzone, editor, *Image and Signal Processing for Remote Sensing XVIII*, volume 8537 of *Proceedings of SPIE*, pages 853704-1–853704-10, 2012.

[16] B. Aiazzi, L. Alparone, S. Baronti, A. Garzelli, and M. Selva. Twenty-five years of pansharpening: A critical review and new developments. In C.-H. Chen, editor, *Signal and Image Processing for Remote Sensing*, pages 533–548. CRC Press, Boca Raton, FL, 2nd edition, 2012.

[17] B. Aiazzi, L. Alparone, S. Baronti, A. Garzelli, and M. Selva. Pan-sharpening of hyperspectral images: A critical analysis of requirements

and assessment on simulated PRISMA data. In L. Bruzzone, editor, *Image and Signal Processing for Remote Sensing XIX*, volume 8892 of *Proceedings of SPIE*, pages 889203–1, 2013.

[18] B. Aiazzi, L. Alparone, S. Baronti, and F. Lotti. Lossless image compression by quantization feedback in a content-driven enhanced Laplacian pyramid. *IEEE Transactions on Image Processing*, 6(6):831–843, June 1997.

[19] B. Aiazzi, L. Alparone, S. Baronti, and I. Pippi. Fusion of 18 m MOMS-2P and 30 m Landsat TM multispectral data by the generalized Laplacian pyramid. *ISPRS International Archives of Photogrammetry and Remote Sensing*, 32(7-4-3W6):116–122, 1999.

[20] B. Aiazzi, L. Alparone, S. Baronti, and I. Pippi. Quality assessment of decision-driven pyramid-based fusion of high resolution multispectral with panchromatic image data. In *IEEE/ISPRS Joint Workshop on Remote Sensing and Data Fusion over Urban Areas*, pages 337–341, 2001.

[21] B. Aiazzi, L. Alparone, S. Baronti, and I. Pippi. Decision-driven pyramid fusion of multi-spectral and panchromatic images. In T. Benes, editor, *Geoinformation for European-wide Integration*, pages 273–278. Millpress, Rotterdam, The Netherlands, 2002.

[22] B. Aiazzi, L. Alparone, S. Baronti, I. Pippi, and M. Selva. Generalised Laplacian pyramid-based fusion of MS + P image data with spectral distortion minimisation. *ISPRS International Archives of Photogrammetry and Remote Sensing*, 34(3B-W3):3–6, 2002.

[23] B. Aiazzi, L. Alparone, S. Baronti, I. Pippi, and M. Selva. Context modeling for joint spectral and radiometric distortion minimization in pyramid-based fusion of MS and P image data. In S. B. Serpico, editor, *Image and Signal Processing for Remote Sensing VIII*, volume 4885 of *Proceedings of SPIE*, pages 46–57, 2003.

[24] B. Aiazzi, L. Alparone, S. Baronti, L. Santurri, and M. Selva. Spatial resolution enhancement of ASTER thermal bands. In L. Bruzzone, editor, *Image and Signal Processing for Remote Sensing XI*, volume 5982 of *Proceedings of SPIE*, pages 59821G–1, 2005.

[25] B. Aiazzi, S. Baronti, F. Lotti, and M. Selva. A comparison between global and context-adaptive pansharpening of multispectral images. *IEEE Geoscience and Remote Sensing Letters*, 6(2):302–306, April 2009.

[26] B. Aiazzi, S. Baronti, and M. Selva. Improving component substitution pansharpening through multivariate regression of MS+Pan data. *IEEE Transactions on Geoscience and Remote Sensing*, 45(10):3230–3239, October 2007.

[27] B. Aiazzi, S. Baronti, M. Selva, and L. Alparone. Bi-cubic interpolation for shift-free pan-sharpening. *ISPRS Journal of Photogrammetry and Remote Sensing*, 86(6):65–76, December 2013.

[28] F. A. Al-Wassai, N. V. Kalyankar, and A. A. Al-Zuky. The IHS transformations based image fusion. *International Journal of Advanced Research in Computer Science*, 2(5), September–October 2011.

[29] L. Alparone, B. Aiazzi, S. Baronti, and A. Garzelli. Fast classified pansharpening with spectral and spatial distortion optimization. In *Proceedings of the IEEE International Geoscience and Remote Sensing Symposium (IGARSS)*, pages 154–157, 2012.

[30] L. Alparone, B. Aiazzi, S. Baronti, A. Garzelli, and F. Nencini. Information-theoretic assessment of fusion of multi-spectral and panchromatic images. In *Proceedings of the 9th IEEE International Conference on Information Fusion*, pages 1–5, 2006.

[31] L. Alparone, B. Aiazzi, S. Baronti, A. Garzelli, F. Nencini, and M. Selva. Spectral information extraction from very high resolution images through multiresolution fusion. In *Image and Signal Processing for Remote Sensing X*, volume 5573 of *Proceedings of SPIE*, pages 1–8, 2004.

[32] L. Alparone, B. Aiazzi, S. Baronti, A. Garzelli, F. Nencini, and M. Selva. Multispectral and panchromatic data fusion assessment without reference. *Photogrammetric Engineering and Remote Sensing*, 74(2):193–200, February 2008.

[33] L. Alparone, S. Baronti, A. Garzelli, and F. Nencini. A global quality measurement of pan-sharpened multispectral imagery. *IEEE Geoscience and Remote Sensing Letters*, 1(4):313–317, October 2004.

[34] L. Alparone, S. Baronti, A. Garzelli, and F. Nencini. Landsat ETM+ and SAR image fusion based on generalized intensity modulation. *IEEE Transactions on Geoscience and Remote Sensing*, 42(12):2832–2839, December 2004.

[35] L. Alparone, V. Cappellini, L. Mortelli, B. Aiazzi, S. Baronti, and R. Carlà. A pyramid-based approach to multisensor image data fusion with preservation of spectral signatures. In P. Gudmandsen, editor, *Future Trends in Remote Sensing*, pages 418–426. Balkema, Rotterdam, The Netherlands, 1998.

[36] L. Alparone, L. Facheris, S. Baronti, A. Garzelli, and F. Nencini. Fusion of multispectral and SAR images by intensity modulation. In *Proceedings of the Seventh International Conference on Information Fusion, FUSION 2004*, volume 2, pages 637–643, 2004.

[37] L. Alparone, L. Wald, J. Chanussot, C. Thomas, P. Gamba, and L. M. Bruce. Comparison of pansharpening algorithms: outcome of the 2006 GRS-S data fusion contest. *IEEE Transactions on Geoscience and Remote Sensing*, 45(10):3012–3021, October 2007.

[38] L. Amolins, Y. Zhang, and P. Dare. Wavelet based image fusion techniques - An introduction, review and comparison. *ISPRS Journal of Photogrammetry and Remote Sensing*, 62(4):249–263, September 2007.

[39] I. Amro, J. Mateos, M. Vega, R. Molina, and A. K. Katsaggelos. A survey of classical methods and new trends in pansharpening of multispectral images. *EURASIP Journal on Advances in Signal Processing*, 2011.

[40] F. Argenti and L. Alparone. Filterbanks design for multisensor data fusion. *IEEE Signal Processing Letters*, 7(5):100–103, May 2000.

[41] F. Argenti and L. Alparone. Speckle removal from SAR images in the undecimated wavelet domain. *IEEE Transactions on Geoscience and Remote Sensing*, 40(11):2363–2374, November 2002.

[42] F. Argenti, A. Lapini, T. Bianchi, and L. Alparone. A tutorial on speckle reduction in synthetic aperture radar images. *IEEE Geoscience and Remote Sensing Magazine*, 1(3):6–35, September 2013.

[43] R. H. Bamberger and M. J. T. Smith. A filter bank for the directional decomposition of images: theory and design. *IEEE Transactions on Signal Processing*, 40(4):882–893, April 1992.

[44] A. Bannari, D. Morin, G. B. Bénié, and F. J. Bonn. A theoretical review of different mathematical models of geometric corrections applied to remote sensing images. *Remote Sensing Reviews*, 13(1–2):27–47, March 1995.

[45] S. Baronti, B. Aiazzi, M. Selva, A. Garzelli, and L. Alparone. A theoretical analysis of the effects of aliasing and misregistration on pansharpened imagery. *IEEE Journal of Selected Topics in Signal Processing*, 5(3):446–453, June 2011.

[46] S. Baronti, A. Casini, F. Lotti, and L. Alparone. Content-driven differential encoding of an enhanced image pyramid. *Signal Processing: Image Communication*, 6(5):463–469, October 1994.

[47] G. Beylkin. On the representation of operators in bases of compactly supported wavelets. *SIAM Journal on Numerical Analysis*, 29(6):1716–1740, December 1992.

[48] F. C. Billingsley. Data processing and reprocessing. In R. N. Colwell, editor, *Manual of Remote Sensing*, volume 1, pages 719–792. American Society of Photogrammetry, Falls Church, 1983.

[49] P. Blanc, T. Blu, T. Ranchin, L. Wald, and R. Aloisi. Using iterated rational filter banks within the ARSIS concept for producing 10 m Landsat multispectral images. *International Journal of Remote Sensing*, 19(12):2331–2343, August 1998.

[50] F. Bovolo, L. Bruzzone, L. Capobianco, S. Marchesi, F. Nencini, and A. Garzelli. Analysis of the effects of pansharpening in change detection on VHR images. *IEEE Geoscience and Remote Sensing Letters*, 7(1):53–57, January 2010.

[51] L. Brown. A survey of image registration techniques. *ACM Computer Surveys*, 24(4):325–376, December 1992.

[52] L. Bruzzone, L. Carlin, L. Alparone, S. Baronti, A. Garzelli, and F. Nencini. Can multiresolution fusion techniques improve classification accuracy? In *Image and Signal Processing for Remote Sensing XII*, volume 6365 of *Proceedings of SPIE*, pages 636509–1, 2006.

[53] V. Buntilov and R. Bretschneider. A content separation image fusion approach: toward conformity between spectral and spatial information. *IEEE Transactions on Geoscience and Remote Sensing*, 45(10):3252–3263, October 2007.

[54] P. J. Burt and E. H. Adelson. The Laplacian pyramid as a compact image code. *IEEE Transactions on Communications*, COM–31(4):532–540, April 1983.

[55] Younggi Byun, Jaewan Choi, and Youkyung Han. An area-based image fusion scheme for the integration of SAR and optical satellite imagery. *IEEE Journal of Selected Topics in Applied Earth Observations and Remote Sensing*, 6(5):2212–2220, October 2013.

[56] R. Caloz and C. Collet. *Précis de télédétection*. Université de Québec, Canada, 2001.

[57] J. B. Campbell and R. H. Wynne. *Introduction to Remote Sensing*. The Guilford Press, New York, NY, 5th edition, 2011.

[58] E. J. Candès and D. L. Donoho. New tight frames of curvelets and optimal representations of objects with piecewise C^2 singularities. *Communications on Pure and Applied Mathematics*, 57(2):219–266, February 2004.

[59] E. J. Candès, J. Romberg, and T. Tao. Robust uncertainty principles: Exact signal reconstruction from highly incomplete frequency information. *IEEE Transactions on Information Theory*, 52(2):489–509, February 2006.

[60] L. Capobianco, A. Garzelli, F. Nencini, L. Alparone, and S. Baronti. Spatial enhancement of Hyperion hyperspectral data through ALI panchromatic image. In *Proceedings of IEEE International Geoscience and Remote Sensing Symposium (IGARSS)*, pages 5158–5161, 2007.

[61] W. Carper, T. Lillesand, and R. Kiefer. The use of intensity-hue-saturation transformations for merging SPOT panchromatic and multi-spectral image data. *Photogrammetric Engineering and Remote Sensing*, 56(4):459–467, April 1990.

[62] E. Catmull and R. Rom. A class of local interpolating splines. In R. E. Barnhill and R. F. Riesenfeld, editors, *Computer Aided Geometric Design*, pages 317–326. Academic Press, New York, NY, 1974.

[63] R. E. Chapman. *Physics for Geologists*. UCL Press, London, UK, 1995.

[64] P. S. Chavez Jr. and J. A. Bowell. Comparison of the spectral information content of Landsat thematic mapper and SPOT for three different sites in the Phoenix Arizona region. *Photogrammetric Engineering and Remote Sensing*, 54(12):1699–1708, December 1988.

[65] P. S. Chavez, Jr. and A. W. Kwarteng. Extracting spectral contrast in Landsat Thematic Mapper image data using selective principal component analysis. *Photogrammetric Engineering and Remote Sensing*, 55(3):339–348, March 1989.

[66] P. S. Chavez, Jr., S. C. Sides, and J. A. Anderson. Comparison of three different methods to merge multiresolution and multispectral data: Landsat TM and SPOT panchromatic. *Photogrammetric Engineering and Remote Sensing*, 57(3):295–303, March 1991.

[67] C.-M. Chen, G. F. Hepner, and R. R. Forster. Fusion of hyperspectral and radar data using the IHS transformation to enhance urban surface features. *ISPRS Journal of Photogrammetry and Remote Sensing*, 58(1–2):19–30, June 2003.

[68] S. Chen, D. L. Donoho, and M. Saunders. Atomic decomposition by basis pursuit. *SIAM Journal on Scientific Computing*, 20(1):33–61, 1998.

[69] X. Chen, Y. Yamaguchi, J. Chen, and Y. Shi. Scale effect of vegetation-index-based spatial sharpening for thermal imagery: A simulation study by ASTER data. *IEEE Geoscience and Remote Sensing Letters*, 9(4):549–553, July 2012.

[70] M. Cheng, C. Wang, and J. Li. Sparse representation based pansharpening using trained dictionary. *IEEE Geoscience and Remote Sensing Letters*, 11(1):293–297, January 2014.

[71] Y. Chibani and A. Houacine. The joint use of IHS transform and redundant wavelet decomposition for fusing multispectral and panchromatic images. *International Journal of Remote Sensing*, 23(18):3821–3833, September 2002.

[72] J. Choi, K. Yu, and Y. Kim. A new adaptive component-substitution-based satellite image fusion by using partial replacement. *IEEE Transactions on Geoscience and Remote Sensing*, 49(1):295–309, January 2011.

[73] H. Chu and W. Zhu. Fusion of IKONOS satellite imagery using IHS transform and local variation. *IEEE Geoscience and Remote Sensing Letters*, 5(4):653–657, October 2008.

[74] A. Cohen, I. Daubechies, and J. C. Feauveau. Biorthogonal bases of compactly supported wavelets. *Communications on Pure and Applied Mathematics*, 45(5):485–560, June 1995.

[75] M. K. Cook, B. A. Peterson, G. Dial, L. Gibson, F. W. Gerlach, K. S. Hutchins, R. Kudola, and H. S. Bowen. IKONOS technical performance assessment. In Sylvia S. Shen and Michael R. Descour, editors, *Algorithms for Multispectral, Hyperspectral, and Ultraspectral Imagery VII*, volume 4381 of *Proceedings of SPIE*, pages 94–108, 2001.

[76] R. E. Crochiere and L. R. Rabiner. *Multirate Digital Signal Processing.* Prentice Hall, Englewood Cliffs, NJ, 1983.

[77] J. C. Curlander and R. N. McDonough. *Synthetic Aperture Radar: Systems and Signal Processing.* Wiley, New York, NY, 1991.

[78] A. L. da Cunha, J. Zhou, and M. N. Do. The nonsubsampled contourlet transform: theory, design, and applications. *IEEE Transactions on Image Processing*, 15(10):3089–3101, October 2006.

[79] P. Dare and I. Dowman. An improved model for automatic feature-based registration of SAR and SPOT images. *ISPRS Journal of Photogrammetry and Remote Sensing*, 56(1):13–28, June 2003.

[80] I. Daubechies. Orthonormal bases of compactly supported wavelets. *Communications on Pure and Applied Mathematics*, 41(7):909–996, October 1988.

[81] I. Daubechies. *Ten Lectures on Wavelets.* SIAM: Society for Industrial and Applied Mathematics, Philadelphia, PA, 1992.

[82] S. de Béthune, F. Muller, and M. Binard. Adaptive intensity matching filters: a new tool for multi-resolution data fusion. In *Multi-sensor systems and data fusion for telecommunications, remote sensing and radar*, pages 1–13. North Atlantic Treaty Organization. Advisory Group for Aerospace Research and Development, 1997.

[83] A. J. De Leeuw, L. M. M. Veugen, and H. T. C. van Stokkom. Geometric correction of remotely-sensed imagery using ground control points and orthogonal polynomials. *International Journal of Remote Sensing*, 9(10–11):1751–1759, October 1988.

[84] A. Della Ventura, A. Rampini, and R. Schettini. Image registration by the recognition of corresponding structures. *IEEE Transactions on Geoscience and Remote Sensing*, 28(3):305–314, May 1990.

[85] M. N. Do. *Directional multiresolution image representations*. PhD thesis, School of Computer and Communication Sciences, Swiss Federal Institute of Technology, Lausanne, Switzerland, 2001.

[86] M. N. Do and M. Vetterli. The finite ridgelet transform for image representation. *IEEE Transactions on Image Processing*, 12(1):16–28, January 2003.

[87] M. N. Do and M. Vetterli. The contourlet transform: an efficient directional multiresolution image representation. *IEEE Transactions on Image Processing*, 14(12):2091–2106, December 2005.

[88] N. A. Dodgson. Image resampling. Technical Report 261, Computer Laboratories, University of Cambridge, Cambridge, UK, August 1992.

[89] N. A. Dodgson. Quadratic interpolation for image resampling. *IEEE Transactions on Image Processing*, 6(9):1322–1326, September 1997.

[90] D. L. Donoho. Compressed sensing. *IEEE Transactions on Information Theory*, 52(4):1289–1306, April 2006.

[91] W. Dou, Y. Chen, X. Li, and D. Sui. A general framework for component substitution image fusion: An implementation using fast image fusion method. *Computers and Geoscience*, 33(2):219–228, February 2007.

[92] Q. Du, N. H. Younan, R. L. King, and V. P. Shah. On the performance evaluation of pan-sharpening techniques. *IEEE Geoscience and Remote Sensing Letters*, 4(4):518–522, October 2007.

[93] P. Dutilleux. An implementation of the "algorithme à trous" to compute the wavelet transform. In J. M. Combes, A. Grossman, and Ph. Tchamitchian, editors, *Wavelets: Time-Frequency Methods and Phase Space*, pages 298–304. Springer, Berlin, 1989.

[94] K. Edwards and P. A. Davis. The use of Intensity-Hue-Saturation transformation for producing color shaded-relief images. *Photogrammetric Engineering and Remote Sensing*, 60(11):1369–1374, November 1994.

[95] M. Ehlers, S. Klonus, P. J. Astrand, and P. Rosso. Multi-sensor image fusion for pansharpening in remote sensing. *International Journal of Image and Data Fusion*, 1(1):25–45, February 2010.

[96] M. T. Eismann and R. C. Hardie. Application of the stochastic mixing model to hyperspectral resolution enhancement. *IEEE Transactions on Geoscience and Remote Sensing*, 42(9):1924–1933, September 2004.

[97] M. T. Eismann and R. C. Hardie. Hyperspectral resolution enhancement using high-resolution multispectral imagery with arbitrary response functions. *IEEE Transactions on Geoscience and Remote Sensing*, 43(3):455–465, March 2005.

[98] D. Fasbender, J. Radoux, and P. Bogaert. Bayesian data fusion for adaptable image pansharpening. *IEEE Transactions on Geoscience and Remote Sensing*, 46(6):1847–1857, June 2008.

[99] D. Fenna. *Cartographic Science: A Compendium of Map Projections, with Derivations*. CRC Press, Boca Raton, FL, 2006.

[100] E. L. Fiume. *The Mathematical Structure of Raster Graphics*. Academic Press, San Diego, CA, 1989.

[101] J. Flusser. An adaptive method for image registration. *Pattern Recognition*, 25(1):45–54, January 1992.

[102] J. Flusser and T. Suk. A moment-based approach to registration of images with affine geometric distortion. *IEEE Transactions on Geoscience and Remote Sensing*, 32(2):382–387, March 1994.

[103] L. M. G. Fonseca and B. S. Manjunath. Registration techniques for multisensor remotely sensed imagery. *Photogrammetric Engineering and Remote Sensing*, 62(9):1049–1056, September 1996.

[104] M. R. B. Forshaw, A. Haskell, P. F. Miller, D. J. Stanley, and J. R. G. Townshend. Spatial resolution of remotely-sensed imagery: a review paper. *International Journal of Remote Sensing*, 4(3):497–520, July 1983.

[105] S. Foucher, G. B. Bénié, and J.-M. Boucher. Multiscale MAP filtering of SAR images. *IEEE Transactions on Image Processing*, 10(1):49–60, January 2001.

[106] B. Garguet-Duport, J. Girel, J.-M. Chassery, and G. Pautou. The use of multiresolution analysis and wavelet transform for merging SPOT Panchromatic and multispectral image data. *Photogrammetric Engineering and Remote Sensing*, 62(9):1057–1066, September 1996.

[107] A. Garzelli. Possibilities and limitations of the use of wavelets in image fusion. In *Proceedings of the IEEE International Geoscience and Remote Sensing Symposium (IGARSS)*, pages 66–68, 2002.

[108] A. Garzelli. Wavelet-based fusion of optical and SAR image data over urban area. In *International Archives of Photogrammetry Remote Sensing and Spatial Information Sciences*, volume 34, pages 59–62, 2002.

[109] A. Garzelli. Pansharpening of multispectral images based on nonlocal parameter optimization. *IEEE Transactions on Geoscience and Remote Sensing*, 53(4):2096–2107, April 2015.

[110] A. Garzelli, G. Benelli, M. Barni, and C. Magini. Improving wavelet-based merging of panchromatic and multispectral images by contextual information. In S. B. Serpico, editor, *Image and Signal Processing for Remote Sensing VI*, volume 4170 of *Proceedings of SPIE, EUROPTO Series*, pages 82–91, 2000.

[111] A. Garzelli, L. Capobianco, L. Alparone, B. Aiazzi, S. Baronti, and M. Selva. Hyperspectral pansharpening based on modulation of pixel spectra. In *Proceedings of the 2nd Workshop on Hyperspectral Image and Signal Processing: Evolution in Remote Sensing (WHISPERS)*, pages 1–4, 2010.

[112] A. Garzelli and F. Nencini. Interband structure modeling for pansharpening of very high resolution multispectral images. *Information Fusion*, 6(3):213–224, September 2005.

[113] A. Garzelli and F. Nencini. Fusion of panchromatic and multispectral images by genetic algorithms. In *Proceedings of the IEEE Geoscience and Remote Sensing Symposium (IGARSS)*, pages 3810–3813, 2006.

[114] A. Garzelli and F. Nencini. PAN-sharpening of very high resolution multispectral images using genetic algorithms. *International Journal of Remote Sensing*, 27(15):3273–3292, August 2006.

[115] A. Garzelli and F. Nencini. Panchromatic sharpening of remote sensing images using a multiscale Kalman filter. *Pattern Recognition*, 40(12):3568–3577, December 2007.

[116] A. Garzelli and F. Nencini. Hypercomplex quality assessment of multi-/hyper-spectral images. *IEEE Geoscience and Remote Sensing Letters*, 6(4):662–665, October 2009.

[117] A. Garzelli, F. Nencini, L. Alparone, B. Aiazzi, and S. Baronti. Pansharpening of multispectral images: A critical review and comparison. In *Proceedings of the IEEE International Geoscience and Remote Sensing Symposium (IGARSS)*, pages 81–84, 2004.

[118] A. Garzelli, F. Nencini, L. Alparone, and S. Baronti. Multiresolution fusion of multispectral and panchromatic images through the curvelet transform. In *Proceedings of the IEEE Geoscience and Remote Sensing Symposium (IGARSS)*, pages 2838–2841, 2005.

[119] A. Garzelli, F. Nencini, and L. Capobianco. Optimal MMSE pan sharpening of very high resolution multispectral images. *IEEE Transactions on Geoscience and Remote Sensing*, 46(1):228–236, January 2008.

[120] A. Garzelli and F. Soldati. Context-driven image fusion of multispectral and panchromatic data based on a redundant wavelet representation. In *IEEE/ISPRS Joint Workshop on Remote Sensing and Data Fusion over Urban Areas*, pages 122–126, 8–9 Nov. 2001.

[121] A. R. Gillespie. Quantifying spatial heterogeneity at the landscape scale using variogram models. *Remote Sensing of Environment*, 42(2):137–145, November 1992.

[122] A. R. Gillespie, A. B. Kahle, and R. E. Walker. Color enhancement of highly correlated images-II. Channel ratio and "Chromaticity" Transform techniques. *Remote Sensing of Environment*, 22(3):343–365, August 1987.

[123] R. C. Gonzalez and R. E. Woods. *Digital image processing*. Prentice Hall, Upple Saddle River, NJ, 3rd edition, 2007.

[124] M. González-Audícana, X. Otazu, O. Fors, and J. Alvarez-Mozos. A low computational-cost method to fuse IKONOS images using the spectral response function of its sensors. *IEEE Transactions on Geoscience and Remote Sensing*, 44(6):1683–1691, June 2006.

[125] M. González-Audícana, X. Otazu, O. Fors, and A. Seco. Comparison between Mallat's and the "à trous" discrete wavelet transform based algorithms for the fusion of multispectral and panchromatic images. *International Journal of Remote Sensing*, 26(3):595–614, February 2005.

[126] M. González-Audícana, J. L. Saleta, R. G. Catalán, and R. García. Fusion of multispectral and panchromatic images using improved IHS and PCA mergers based on wavelet decomposition. *IEEE Transactions on Geoscience and Remote Sensing*, 42(6):1291–1299, June 2004.

[127] A. Goshtasby. Registration of images with geometric distortions. *IEEE Transactions on Geoscience and Remote Sensing*, 26(1):60–64, January 1988.

[128] E. W. Grafarend and F. W. Krumm. *Map Projections: Cartographic Information Systems*. Springer-Verlag, Berlin, Germany, 2006.

[129] H. N. Gross and J. R. Schott. Application of spectral mixture analysis and image fusion techniques for image sharpening. *Remote Sensing of Environment*, 63(2):85–94, February 1998.

[130] R. C. Hardie, M. T. Eismann, and G. L. Wilson. MAP estimation of hyperspectral image resolution enhancement using an auxiliary sensor.

IEEE Transactions on Image Processing, 13(9):1174–1184, September 2004.

[131] J. R. Hardy. Methods and accuracy on location of Landsat MSS points on maps. *Journal of the British Interplanetary Society*, 31(8):305–311, August 1978.

[132] B. A. Harrison and D. L. B. Jupp. *Introduction to Image Processing*. CSRIO Publishing, Melbourne, Australia, 1990.

[133] F. Henderson and A. Lewis. *Manual of Remote Sensing: Principles and Applications of Imaging Radar*. Wiley, New York, NY, 1998.

[134] K. Hoffman and R. Kunze. *Linear Algebra*. Prentice–Hall, Englewood Cliffs, NJ, 2nd edition, 1971.

[135] G. C. Holst. *Sampling, aliasing, and data fidelity for electronic imaging systems, communications, and data acquisition*. JCD Publishing, Winter Park, FL, 1998.

[136] H. Hotelling. Analysis of a complex of statistical variables into Principal Components. *Journal of Educational Psychology*, 24(6):417–441, September 1933.

[137] B. Huang and H. Song. Spatiotemporal reflectance fusion via sparse representation. *IEEE Transactions on Geoscience and Remote Sensing*, 50(10):3707–3716, October 2012.

[138] B. Huang, H. Song, H. Cui, J. Peng, and Z. Xu. Spatial and spectral image fusion using sparse matrix factorization. *IEEE Transactions on Geoscience and Remote Sensing*, 52(3):1693–1704, March 2014.

[139] J. Inglada and A. Giros. On the possibility of automatic multisensor image registration. *IEEE Transactions on Geoscience and Remote Sensing*, 42(10):2104–2120, October 2004.

[140] S. Ioannidou and V. Karathanassi. Investigation of the dual-tree complex and shift-invariant discrete wavelet transforms on Quickbird image fusion. *IEEE Geoscience and Remote Sensing Letters*, 4(1):166–170, January 2007.

[141] A. K. Jain. *Fundamentals of Digital Image Processing*. Prentice Hall, Englewood Cliffs, NJ, 1989.

[142] C. Jeganathan, N. A. S. Hamm, S. Mukherjee, P. M. Atkinson, P. L. N. Raju, and V. K. Dadhwal. Evaluating a thermal image sharpening model over a mixed agricultural landscape in India. *International Journal of Applied Earth Observation and Geoinformation*, 13(2):178–191, April 2011.

[143] Linhai Jing and Qiuming Cheng. An image fusion method taking into account phenological analogies and haze. *International Journal of Remote Sensing*, 32(6):1675–1694, March 2011.

[144] D. B. Judd and G. Wyszecki. *Color in Business, Science and Industry.* Wiley, New York, NY, 3rd edition, 1975.

[145] X. Kang, S. Li, and J. A. Benediktsson. Pansharpening with matting model. *IEEE Transactions on Geoscience and Remote Sensing*, 52(8):5088–5099, August 2014.

[146] M. M. Khan, L. Alparone, and J. Chanussot. Pansharpening based on QNR optimization. In *Proceedings of the IEEE International Geoscience and Remote Sensing Symposium (IGARSS)*, volume 5, pages 73–76, 2008.

[147] M. M. Khan, L. Alparone, and J. Chanussot. Pansharpening quality assessment using modulation transfer function filters. In *Proceedings of the IEEE International Geoscience and Remote Sensing Symposium (IGARSS)*, volume 5, pages 61–64, 2008.

[148] M. M. Khan, L. Alparone, and J. Chanussot. Pansharpening quality assessment using the modulation transfer functions of instruments. *IEEE Transactions on Geoscience and Remote Sensing*, 47(11):3880–3891, November 2009.

[149] M. M. Khan, J. Chanussot, and L. Alparone. Hyperspectral pansharpening using QNR optimization constraint. In *Proceedings of the 1st Workshop on Hyperspectral Image and Signal Processing: Evolution in Remote Sensing (WHISPERS)*, pages 1–4, 2009.

[150] M. M. Khan, J. Chanussot, and L. Alparone. Pansharpening of hyperspectral images using spatial distortion optimization. In *Proceedings of the 16th IEEE International Conference on Image Processing (ICIP)*, pages 2853–2856, 2009.

[151] M. M. Khan, J. Chanussot, L. Condat, and A. Montavert. Indusion: Fusion of multispectral and panchromatic images using the induction scaling technique. *IEEE Geoscience and Remote Sensing Letters*, 5(1):98–102, January 2008.

[152] A. Khandelwal and K. S. Rajan. Hyperspectral image enhancement based on sensor simulation and vector decomposition. In *Proceedings of 14th International Conference on Information Fusion - Fusion 2011*, pages 1234–1239, 2011.

[153] M. G. Kim, I. Dinstein, and L. Shaw. A prototype filter design approach to pyramid generation. *IEEE Transactions on Pattern Analysis and Machine Intelligence*, 15(12):1233–1240, December 1993.

[154] F. A. Kruse and G. L. Raines. A technique for enhancing digital colour images by contrast stretching in Munsell colour space. In *Proceedings of the International Symposium on Remote Sensing of Environment, Third Thematic Conference: Remote Sensing for Exploration Geology*, pages 755–760, 1984.

[155] D. Kundur and D. Hatzinakos. Blind image deconvolution. *IEEE Signal Processing Magazine*, 13(3):43–64, 1996.

[156] W. P. Kustas, J. M. Norman, M. C. Anderson, and A. N. French. Estimating subpixel surface temperatures and energy fluxes from the vegetation indexradiometric temperature relationship. *Remote Sensing of Environment*, 85(4):429–440, June 2003.

[157] C. A. Laben and B. V. Brower. Process for enhancing the spatial resolution of multispectral imagery using pan-sharpening, 2000. U.S. Patent # 6,011,875.

[158] F. Laporterie-Déjean, H. de Boissezon, G. Flouzat, and M.-J. Lefévre-Fonollosa. Thematic and statistical evaluations of five panchromatic/multispectral fusion methods on simulated PLEIADES-HR images. *Information Fusion*, 6(3):193–212, September 2005.

[159] C. Latry, H. Vadon, M. J. Lefevre, and H. De Boissezon. SPOT5 THX: a 2.5m fused product. In *Proceedings 2nd GRSS/ISPRS Joint Workshop on Remote Sensing and Data Fusion over Urban Areas*, pages 87–89, May 2003.

[160] J. Le Moigne. Parallel registration of multi-sensor remotely sensed imagery using wavelet coefficients. In H. Szu, editor, *Wavelet Applications*, volume 2242 of *Proceedings of SPIE*, pages 432–443, 1994.

[161] J. Le Moigne, N. S. Netanyahu, and R. D. Eastman, editors. *Image Registration for Remote Sensing*. Cambridge University Press, Cambridge, UK, 2011.

[162] R. S. Ledley, M. Buas, and T. J. Golab. Fundamentals of true-color image processing. In *Proceedings of the 10th International Conference on Pattern Recognition*, volume 1, pages 791–795, 1990.

[163] S. Leprince, S. Barbot, F. Ayoub, and J.-P. Avouac. Automatic and precise orthorectification, coregistration, and subpixel correlation of satellite images: application to ground deformation measurements. *IEEE Transactions on Geoscience and Remote Sensing*, 45(6):1529–1558, June 2007.

[164] H. Li, B. S. Manjunath, and S. K. Mitra. A contour-based approach to multisensor image registration. *IEEE Transactions on Image Processing*, 4(3):320–334, March 1995.

[165] H. Li, B. S. Manjunath, and S. K. Mitra. Multisensor image fusion using the wavelet transform. *Graphical Models Image Processing*, 57(3):235–245, May 1995.

[166] S. Li, J. T. Kwok, and Y. Wang. Using the discrete wavelet frame transform to merge Landsat TM and SPOT panchromatic images. *Information Fusion*, 3(1):17–23, March 2002.

[167] S. Li and B. Yang. A new pan-sharpening method using a compressed sensing technique. *IEEE Transactions on Geoscience and Remote Sensing*, 49(2):738–746, February 2011.

[168] S. Li, H. Yin, and L. Fang. Remote sensing image fusion via sparse representations over learned dictionaries. *IEEE Transactions on Geoscience and Remote Sensing*, 51(9):4779–4789, September 2013.

[169] Shuang Li, Zhilin Li, and Jianya Gonga. Multivariate statistical analysis of measures for assessing the quality of image fusion. *International Journal of Image and Data Fusion*, 1(1):47–66, March 2010.

[170] Z. Li and H. Leung. Fusion of multispectral and panchromatic images using a restoration-based method. *IEEE Transactions on Geoscience and Remote Sensing*, 47(5):1482–1491, March 2009.

[171] G. A. Licciardi, M. M. Khan, J. Chanussot, A. Montanvert, L. Condat, and C. Jutten. Fusion of hyperspectral and panchromatic images using multiresolution analysis and nonlinear PCA band reduction. *EURASIP Journal on Advances in Signal Processing*, 2012(1):207:1–207:17, September 2012.

[172] T. M. Lillesand and R. W. Kiefer. *Remote Sensing and Image Interpretation*. Wiley, New York, NY, 2nd edition, 1987.

[173] Y. Ling, M. Ehlers, E. Lynn Usery, and M. Madden. FFT-enhanced IHS transform method for fusing high-resolution satellite images. *ISPRS Journal of Photogrammetry and Remote Sensing*, 61(6):381–392, February 2007.

[174] J. G. Liu. Smoothing filter based intensity modulation: a spectral preserve image fusion technique for improving spatial details. *International Journal of Remote Sensing*, 21(18):3461–3472, December 2000.

[175] N. T. Lloyd and D. Bau. *Numerical Linear Algebra*. SIAM: Society for Industrial and Applied Mathematics, Philadelphia, PA, 3rd edition, 1997.

[176] S. P. Lloyd. Least squares quantization in PCM. *IEEE Transactions on Information Theory*, 28(2):129–137, March 1982.

[177] E. Maeland. On the comparison of the interpolation methods. *IEEE Transactions on Medical Imaging*, 7(3):213–217, September 1988.

[178] S. Mallat. A theory for multiresolution signal decomposition: the wavelet representation. *IEEE Transactions on Pattern Analysis and Machine Intelligence*, 11(7):674–693, July 1989.

[179] S. G. Mallat. *A Wavelet Tour of Signal Processing*. Academic Press, New York, 2nd edition, 1999.

[180] J. V. Martonchik, C. J. Bruegge, and A. Strahler. A review of reflectance nomenclature used in remote sensing. *Remote Sensing Reviews*, 19(1–4):9–20, December 2000.

[181] P. M. Mather and M. Koch. *Computer Processing of Remotely-Sensed Images: An Introduction*. Wiley-Blackwell, Oxford, UK, 4th edition, 2011.

[182] S. Max. Quantizing for minimum distortion. *IRE Transactions on Information Theory*, 6(1):7–12, March 1960.

[183] M. F. McCabe, L. K. Balik, J. Theiler, and A. R. Gillespie. Linear mixing in thermal infrared temperature retrieval. *International Journal of Remote Sensing*, 29(17–18):5047–5061, September 2008.

[184] H. McNairn, C. Champagne, J. Shang, D. Holmstrom, and G. Reichert. Integration of optical and synthetic aperture radar (SAR) imagery for delivering operational annual crop inventories. *ISPRS Journal of Photogrammetry and Remote Sensing*, 64(5):434–449, September 2009.

[185] E. H. W. Meijering, K. J. Zuiderveld, and M. A. Viergever. Image reconstruction by convolution with symmetrical piecewise nth-order polynomial kernels. *IEEE Transactions on Image Processing*, 8(2):192–201, February 1999.

[186] R. A. Meyer and C. S. Burrus. Unified analysis of multirate and periodically time-varying digital filters. *IEEE Transactions on Circuits and Systems*, CAS-22(3):162–168, March 1975.

[187] F. A. Mianji, Ye Zhang, Yanfeng Gu, and A. Babakhani. Spatial-spectral data fusion for resolution enhancement of hyperspectral imagery. In *Proceedings of IEEE International Geoscience and Remote Sensing Symposium (IGARSS)*, pages 1011–1014, 2009.

[188] D. P. Mitchell and A. N. Netravali. Reconstruction filters in computer graphics. *Computer Graphics*, 22(4):221–228, August 1988.

[189] M. Moeller. Remote sensing for the monitoring of urban growth patterns. In M. Moeller and E. Wentz, editors, *Remote Sensing and Spatial Information Sciences*, volume XXXVI - 8/W27 of *The International Archives of the Photogrammetry*, pages 0–6, 2005.

[190] M. Moghaddam, J. L. Dungan, and S. Acker. Forest variable estimation from fusion of SAR and multispectral optical data. *IEEE Transactions on Geoscience and Remote Sensing*, 40(10):2176–2187, October 2002.

[191] A. Mohammadzadeh, A. Tavakoli, and M. J. Valadan Zoej. Road extraction based on fuzzy logic and mathematical morphology from pan-sharpened IKONOS images. *The Photogrammetric Record*, 21(113):44–60, February 2006.

[192] M. S. Moran. A window-based technique for combining Landsat Thematic Mapper thermal data with higher-resolution multispectral data over agricultural lands. *Photogrammetric Engineering and Remote Sensing*, 56(3):337–342, March 1990.

[193] A. Moreira, P. Prats-Iraola, M. Younis, G. Krieger, I. Hajnsek, and K. P. Papathanassiou. A tutorial on synthetic aperture radar. *IEEE Geoscience and Remote Sensing Magazine*, 1(1):6–43, March 2013.

[194] G. P. Nason and B. W. Silverman. The stationary wavelet transform and some statistical applications. In A. Antoniadis and G. Oppenheim, editors, *Wavelets and Statistics*, volume 103, pages 281–299. Springer-Verlag, New York, NY, USA, 1995.

[195] F. Nencini, A. Garzelli, S. Baronti, and L. Alparone. Remote sensing image fusion using the curvelet transform. *Information Fusion*, 8(2):143–156, April 2007.

[196] J. Núñez, X. Otazu, O. Fors, A. Prades, V. Palà, and R. Arbiol. Multiresolution-based image fusion with additive wavelet decomposition. *IEEE Transactions on Geoscience and Remote Sensing*, 37(3):1204–1211, May 1999.

[197] H. Nyquist. Certain topics in telegraph transmission theory. *Transactions of the AIEE*, 47(2):617–644, April 1928.

[198] C. Oliver and S. Quegan. *Understanding Synthetic Aperture Radar Images*. SciTech Publishing, Herndon, VA, 2nd edition, 2004.

[199] A. V. Oppenheim and R. W. Schafer. *Discrete-Time Signal Processing*. Prentice Hall, Englewood Cliffs, NJ, 1989.

[200] X. Otazu, M. González-Audícana, O. Fors, and J. Núñez. Introduction of sensor spectral response into image fusion methods. Application to wavelet-based methods. *IEEE Transactions on Geoscience and Remote Sensing*, 43(10):2376–2385, October 2005.

[201] K. Ouchi. Recent trend and advance of synthetic aperture radar with selected topics. *Remote Sensing*, 5(2):716–807, February 2013.

[202] G. Pajares and J. M. de la Cruz. A wavelet-based image fusion tutorial. *Pattern Recognition*, 37(9):1855–1872, September 2004.

[203] F. Palsson, J. R. Sveinsson, and M. O. Ulfarsson. A new pansharpening algorithm based on total variation. *IEEE Geoscience and Remote Sensing Letters*, 11(1):318–322, January 2014.

[204] Z. Pan, J. Yu, H. Huang, S. Hu, A. Zhang, H. Ma, and W. Sun. Super-resolution based on compressive sensing and structural self-similarity for remote sensing images. *IEEE Transactions on Geoscience and Remote Sensing*, 51(9):4864–4876, September 2013.

[205] S. C. Park, M. K. Park, and M. G. Kang. Super-resolution image reconstruction: a technical overview. *IEEE Signal Processing Magazine*, 20(3):21–36, May 2003.

[206] S. K. Park and R. A. Schowengerdt. Image reconstruction by parametric cubic convolution. *Computer Vision, Graphics, and Image Processing*, 23(3):258–272, September 1983.

[207] J. A. Parker, R. V. Kenyon, and D. E. Troxel. Comparison of interpolation methods for image resampling. *IEEE Transactions on Medical Imaging*, MI-2(1):31–39, March 1983.

[208] Honghong Peng and Raghuveer Rao. Hyperspectral image enhancement with vector bilateral filtering. In *Proceedings of the International Conference on Image Processing (ICIP)*, volume 2, pages 3713–3716, 2009.

[209] B. Penna, T. Tillo, E. Magli, and G. Olmo. Transform coding techniques for lossy hyperspectral data compression. *IEEE Transactions on Geoscience and Remote Sensing*, 45(5):1408–1421, May 2007.

[210] G. Piella and H. Heijmans. A new quality metric for image fusion. In *Proceedings of the International Conference on Image Processing (ICIP)*, volume 2, pages III – 173–176, 2003.

[211] D. D. Y. Po and M. N. Do. Directional multiscale modeling of images using the contourlet transform. *IEEE Transactions on Image Processing*, 15(6):1610–1620, June 2006.

[212] C. Pohl and J. L. van Genderen. Multisensor image fusion in remote sensing: concepts, methods and applications. *International Journal of Remote Sensing*, 19(5):823–854, March 1998.

[213] V. Poulain, J. Inglada, M. Spigai, J. Y. Tourneret, and P. Marthon. High-resolution optical and SAR image fusion for building database updating. *IEEE Transactions on Geoscience and Remote Sensing*, 49(8):2900–2910, August 2011.

[214] P. S. Pradhan, R. L. King, N. H. Younan, and D. W. Holcomb. Estimation of the number of decomposition levels for a wavelet-based multiresolution multisensor image fusion. *IEEE Transactions on Geoscience and Remote Sensing*, 44(12):3674–3686, December 2006.

[215] J. C. Price. Combining panchromatic and multispectral imagery from dual resolution satellite instruments. *Remote Sensing of Environment*, 21(2):119–128, March 1987.

[216] T. Ranchin, B. Aiazzi, L. Alparone, S. Baronti, and L. Wald. Image fusion - the ARSIS concept and some successful implementation schemes. *ISPRS Journal of Photogrammetry and Remote Sensing*, 58(1-2):4–18, June 2003.

[217] T. Ranchin and L. Wald. Fusion of high spatial and spectral resolution images: the ARSIS concept and its implementation. *Photogrammetric Engineering and Remote Sensing*, 66(1):49–61, January 2000.

[218] M. K. Rangaswamy. Quickbird II two-dimensional on-orbit Modulation Transfer Function analysis using convex mirror array. Master's thesis, South Dakota State University, 2003.

[219] W. G. Rees. *Physical Principles of Remote Sensing*. Cambridge University Press, Cambridge, UK, 2nd edition, 2001.

[220] J. A. Richards. *Remote Sensing with Imaging Radar*. Springer, Heidelberg, Germany, 2009.

[221] J. A. Richards and X. Jia. *Remote Sensing Digital Image Analysis, An Introduction*. Springer–Verlag, Berlin–Heidelberg, Germany, 5th edition, 2012.

[222] F. F. Sabins. *Remote Sensing: Principles and Interpretations*. W. H. Freeman, New York, NY, 1996.

[223] R. W. Schafer and L. R. Rabiner. A digital signal processing approach to interpolation. *Proceedings of the IEEE*, 61(6):692–702, June 1973.

[224] A. H. Schistad Solberg, A. K. Jain, and T. Taxt. Multisource classification of remotely sensed data: fusion of Landsat TM and SAR images. *IEEE Transactions on Geoscience and Remote Sensing*, 32(4):768–778, July 1994.

[225] A. H. Schistad Solberg, T. Taxt, and A. K. Jain. A Markov random field model for classification of multisource satellite imagery. *IEEE Transactions on Geoscience and Remote Sensing*, 34(1):100–113, January 1996.

[226] R. A. Schowengerdt. *Remote Sensing: Models and Methods for Image Processing*. Academic Press, Orlando, FL, USA, 2nd edition, 1997.

[227] R. A. Schowengerdt. *Remote Sensing: Models and Methods for Image Processing.* Elsevier, The Netherlands, 3rd edition, 2007.

[228] M. Selva, B. Aiazzi, F. Butera, L. Chiarantini, and S. Baronti. Hypersharpening of hyperspectral data: a first approach. In *Proceedings of the 6th Workshop on Hyperspectral Image and Signal Processing: Evolution in Remote Sensing (WHISPERS)*, pages 1–4, 2014.

[229] V. P. Shah, N. H. Younan, and R. L. King. An efficient pan-sharpening method via a combined adaptive-PCA approach and contourlets. *IEEE Transactions on Geoscience and Remote Sensing*, 46(5):1323–1335, May 2008.

[230] C. E. Shannon. Communication in the presence of noise. *Proceedings of the Institute of Radio Engineers*, 37(1):10–21, January 1949.

[231] M. J. Shensa. The discrete wavelet transform: wedding the à trous and Mallat algorithm. *IEEE Transactions on Signal Processing*, 40(10):2464–2482, October 1992.

[232] V. K. Shettigara. A generalized component substitution technique for spatial enhancement of multispectral images using a higher resolution data set. *Photogrammetric Engineering and Remote Sensing*, 58(5):561–567, May 1992.

[233] J. Shi and S. E. Reichenbach. Image interpolation by two-dimensional parametric cubic convolution. *IEEE Transactions on Image Processing*, 15(7):1857–1870, July 2006.

[234] D. S. Simonett. The development and principles of remote sensing. In R. N. Colwell, editor, *Manual of Remote Sensing*, volume 1, pages 1–36. American Society of Photogrammetry, Falls Church, VA, 1983.

[235] P. Sirguey, R. Mathieu, Y. Arnaud, M. M. Khan, and J. Chanussot. Improving MODIS spatial resolution for snow mapping using wavelet fusion and ARSIS concept. *IEEE Geoscience and Remote Sensing Letters*, 5(1):78–82, January 2008.

[236] P. N. Slater. *Remote Sensing, Optics and Optical Systems.* Addison-Wesley, Reading, MA, 1980.

[237] A. R. Smith. Color gamut transform pairs. *Computer Graphics*, 12(3):12–19, August 1978.

[238] C. Smith. *Environmental Physics.* Routledge, London, UK, 2001.

[239] H. Song and B. Huang. Spatiotemporal satellite image fusion through one-pair image learning. *IEEE Transactions on Geoscience and Remote Sensing*, 51(4):1883–1896, 2013.

[240] M. Sonka, V. Klavac, and R. Boyle. *Image Processing, Analysis, and Machine Vision.* Thomson Learning, London, UK, 3rd edition, 2007.

[241] C. Souza Jr, L. Firestone, L. M. Silva, and D. Roberts. Mapping forest degradation in the Eastern Amazon from SPOT 4 through spectral mixture models. *Remote Sensing of Environment*, 87(4):494–506, November 2003.

[242] J. L. Starck, E. J. Candes, and D. L. Donoho. The curvelet transform for image denoising. *IEEE Transactions on Image Processing*, 11(6):670–684, June 2002.

[243] J.-L. Starck, J. Fadili, and F. Murtagh. The undecimated wavelet decomposition and its reconstruction. *IEEE Transactions on Image Processing*, 16(2):297–309, February 2007.

[244] G. Strang and T. Nguyen. *Wavelets and Filter Banks.* Wellesley Cambridge Press, Wellesley, MA, 2nd edition, 1996.

[245] O. Thépaut, K. Kpalma, and J. Ronsin. Automatic registration of ERS and SPOT multisensor images in a data fusion context. *Forest Ecology and Management*, 128(1-2):93–100, March 2000.

[246] C. Thomas, T. Ranchin, L. Wald, and J. Chanussot. Synthesis of multispectral images to high spatial resolution: a critical review of fusion methods based on remote sensing physics. *IEEE Transactions on Geoscience and Remote Sensing*, 46(5):1301–1312, May 2008.

[247] K. Tomiyasu. Tutorial review of synthetic-aperture radar (SAR) with applications to imaging of the ocean surface. *Proceedings of the IEEE*, 66(5):563–583, January 1978.

[248] T. Toutin. Intégration de données multi-sources: comparaison de méthodes géometriques et radiométriques. *International Journal of Remote Sensing*, 16(15):2795–2811, October 1995.

[249] T. Toutin. Review article: Geometric processing of remote sensing images: models, algorithms and methods. *International Journal of Remote Sensing*, 25(10):1893–1924, May 2004.

[250] J. R. G. Townshend. The spatial resolving power of Earth resources satellites: A review. Technical Report 82020, NASA Goddard Space Flight Center, Greenbelt, MD, May 1980.

[251] T.-M. Tu, P. S. Huang, C.-L. Hung, and C.-P. Chang. A fast intensity-hue-saturation fusion technique with spectral adjustment for IKONOS imagery. *IEEE Geoscience and Remote Sensing Letters*, 1(4):309–312, October 2004.

[252] T.-M. Tu, S.-C. Su, H.-C. Shyu, and P. S. Huang. A new look at IHS-like image fusion methods. *Information Fusion*, 2(3):177–186, September 2001.

[253] F. T. Ulaby, F. Kouyate, B. Brisco, and T. H. L. Williams. Textural information in SAR images. *IEEE Transactions on Geoscience and Remote Sensing*, 24(2):235–245, March 1986.

[254] T. Updike and C. Comp. Radiometric use of WorldView-2 imagery. Technical report, DigitalGlobe, Longmont, CO, November 2010.

[255] P. P. Vaidyanathan. *Multirate Systems and Filter Banks*. Prentice Hall, Englewood Cliffs, NJ, 1992.

[256] M. Vetterli and J. Kovacevic. *Wavelets and Subband Coding*. Prentice Hall, Englewood Cliffs, NJ, 1995.

[257] G. Vivone, L. Alparone, J. Chanussot, M. Dalla Mura, A. Garzelli, R. Restaino, G. Licciardi, and L. Wald. A critical comparison among pansharpening algorithms. *IEEE Transactions on Geoscience and Remote Sensing*, 53(5):2565–2586, May 2015.

[258] G. Vivone, R. Restaino, M. Dalla Mura, G. Licciardi, and J. Chanussot. Contrast and error-based fusion schemes for multispectral image pan-sharpening. *IEEE Geoscience and Remote Sensing Letters*, 11(5):930–934, May 2014.

[259] J. Vrabel. Multispectral imagery band sharpening study. *Photogrammetric Engineering and Remote Sensing*, 62(9):1075–1086, September 1996.

[260] L. Wald. *Data Fusion: Definitions and Architectures — Fusion of Images of Different Spatial Resolutions*. Les Presses del l'École des Mines, Paris, France, 2002.

[261] L. Wald, T. Ranchin, and M. Mangolini. Fusion of satellite images of different spatial resolutions: assessing the quality of resulting images. *Photogrammetric Engineering and Remote Sensing*, 63(6):691–699, June 1997.

[262] Z. Wang and A. C. Bovik. A universal image quality index. *IEEE Signal Processing Letters*, 9(3):81–84, March 2002.

[263] Z. Wang and A. C. Bovik. Mean squared error: Love it or leave it? A new look at signal fidelity measures. *IEEE Signal Processing Magazine*, 26(1):98–117, January 2009.

[264] Z. Wang, D. Ziou, C. Armenakis, D. Li, and Q. Li. A comparative analysis of image fusion methods. *IEEE Transactions on Geoscience and Remote Sensing*, 43(6):1391–1402, June 2005.

[265] U. Wegmuller, C. L. Werner, D. Nuesch, and M. Borgeaud. Radargrammetry and space triangulation for DEM generation and image orthorectification. In *Proceedings of the IEEE International Geoscience And Remote Sensing Symposium (IGARSS)*, volume 1, pages 179–181, 2003.

[266] G. Wolberg. *Digital Image Warping*. IEEE Computer Society Press, Los Alamitos, CA, 1990.

[267] K. W. Wong. Geometric and cartography accuracy of ERTS-1 imagery. *Photogrammetric Engineering and Remote Sensing*, 41(5):621–635, May 1975.

[268] H. Xie, L. E. Pierce, and F. T. Ulaby. Statistical properties of logarithmically transformed speckle. *IEEE Transactions on Geoscience and Remote Sensing*, 40(3):721–727, March 2002.

[269] J. Xu, Z. Guan, and J. Liu. An improved fusion method for merging multi-spectral and panchromatic images considering sensor spectral response. In Jun Chen, Jie Jiang, and J. van Genderen, editors, *Silk Road for Information from Imagery*, volume XXXVII, Part B7 of *The International Archives of the Photogrammetry, Remote Sensing and Spatial Information Sciences*, pages 1169–1174, 2008.

[270] J. Yang, J. Wright, T.S. Huang, and Y. Ma. Image super-resolution via sparse representation. *IEEE Transactions on Image Processing*, 19(11):2861–2873, November 2010.

[271] D. A. Yocky. Artifacts in wavelet image merging. *Optical Engineering*, 35(7):2094–2101, July 1996.

[272] D. A. Yocky. Multiresolution wavelet decomposition image merger of Landsat Thematic Mapper and SPOT panchromatic data. *Photogrammetric Engineering and Remote Sensing*, 62(9):1067–1074, September 1996.

[273] R. H. Yuhas, A. F. H. Goetz, and J. W. Boardman. Discrimination among semi-arid landscape endmembers using the Spectral Angle Mapper (SAM) algorithm. In *Proceeding Summaries 3rd Annual JPL Airborne Geoscience Workshop*, pages 147–149, 1992.

[274] W. Zhan, Y. Chen, J. Zhou, J. Wang, W. Liu, J. Voogt, X. Zhu, J. Quan, and J. Li. Disaggregation of remotely sensed land surface temperature: Literature survey, taxonomy, issues, and caveats. *Remote Sensing of Environment*, 131:119–139, April 2013.

[275] Y. Zhang. A new automatic approach for effectively fusing Landsat 7 as well as IKONOS images. In *Proceedings IEEE International Geoscience and Remote Sensing Symposium (IGARSS)*, volume 4, pages 2429–2431, 2002.

[276] Y. Zhang, S. De Backer, and P. Scheunders. Noise-resistant wavelet-based Bayesian fusion of multispectral and hyperspectral images. *IEEE Transactions on Geoscience and Remote Sensing*, 47(11):3834–3843, November 2009.

[277] Y. Zhang, A. Duijster, and P. Scheunders. A Bayesian restoration approach for hyperspectral images. *IEEE Transactions on Geoscience and Remote Sensing*, 50(9):3453–3462, September 2012.

[278] J. Zhou, D. L. Civco, and J. A. Silander. A wavelet transform method to merge Landsat TM and SPOT panchromatic data. *International Journal of Remote Sensing*, 19(4):743–757, March 1998.

[279] X. X. Zhu and R. Bamler. A sparse image fusion algorithm with application to pan-sharpening. *IEEE Transactions on Geoscience and Remote Sensing*, 51(5):2827–2836, May 2013.

[280] B. Zitová and J. Flusser. Image registration methods: a survey. *Image and Vision Computing*, 21(11):977–1000, October 2003.

Index

Printed and bound by CPI Group (UK) Ltd, Croydon, CR0 4YY

01/11/2024

01782617-0009